MICROBES AND EVOLUTION

The World That Darwin Never Saw

MICROBES AND EVOLUTION

The World That Darwin Never Saw

Edited by

Roberto Kolter
Harvard Medical School, Boston, MA 02115

and
Stanley Maloy
San Diego State University, San Diego, CA 92182-1010

ASM Press
Washington, DC 20036

Library of Congress Cataloging-in-Publication Data

Microbes and evolution : the world that Darwin never saw / edited by Roberto Kolter, Stanley Maloy.

p. ; cm.

Includes bibliographical references and index.

ISBN 978-1-55581-540-0 — ISBN 978-1-55581-847-0 (e-ISBN) 1. Microorganisms—Evolution. 2. Bacteria—Evolution. 3. Evolution (Biology) I. Kolter, Roberto, 1953- II. Maloy, Stanley R.

[DNLM: 1. Darwin, Charles, 1809-1882. 2. Biological Evolution—Essays. 3. Microbiological Phenomena—Essays. QH 366.2]

QR13.M522 2012

571.8'92—dc23

2012013688

doi:10.1128/9781555818470

Printed in the United States of America

10 9 8 7 6 5 4 3

Address editorial correspondence to: ASM Press, 1752 N St., N.W., Washington, DC 20036-2904, USA.

Send orders to: ASM Press, P.O. Box 605, Herndon, VA 20172, USA.

Phone: 800-546-2416; 703-661-1593. Fax: 703-661-1501.

E-mail: books@asmusa.org

Online: http://estore.asm.org

Contents

Contributors *ix*

Preface *xiii*

Introduction **Darwin and Microbiology** **1**
Roberto Kolter and Stanley Maloy

Chapter 1 **Evolution in Action: A 50,000-Generation Salute to Charles Darwin** **9**
Richard E. Lenski

Chapter 2 **Minimal Genomes and Reducible Complexity** **17**
Andrés Moya

Chapter 3 **Lady Lumps's Mouthguard** **25**
Jessica Green

Chapter 4 **Trying To Make Sense of the Microbial Census** **31**
Mitchell L. Sogin

Chapter 5 **The View from Below** **37**
Margaret Riley and Robert Dorit

Chapter 6 **Running Wild with Antibiotics** **43**
Roberto Kolter

Chapter 7 **Antibiotic Resistance** **49**
Diarmaid Hughes

Chapter 8 **Bacteria Battling for Survival** **59**
Thomas M. Schmidt

Chapter 9 **Phage: An Important Evolutionary Force Darwin Never Knew** 65
Forest Rohwer

Chapter 10 **The Struggle for Existence: Mutualism** 71
Paul E. Turner

Chapter 11 **The Secret Social Lives of Microorganisms** 77
Kevin R. Foster

Chapter 12 **Microbes and Microevolution** 85
Evgeni Sokurenko

Chapter 13 **Unnecessary Baggage** 93
Stanley Maloy and Guido Mora

Chapter 14 **Bacterial Adaptation: Built-In Responses and Random Variations** 99
Josep Casadesús

Chapter 15 **The Impact of Differential Regulation on Bacterial Speciation** 109
Eduardo A. Groisman

Chapter 16 **An Accidental Evolutionary Biologist: GASP, Long-Term Survival, and Evolution** 115
Steven E. Finkel

Chapter 17 **How Bacteria Revealed Darwin's Mistake (and Got Me To Read *On the Origin of Species*)** 123
John R. Roth

Chapter 18 **The Role of Conjugation in the Evolution of Bacteria** 133
Fernando de la Cruz

Chapter 19 **Do Bacteria Have Sex?** 139
Rosemary J. Redfield

Chapter 20 **Better than Sex** 145
Harald Brüssow

Chapter 21 **Darwin in My Lab: Mutation, Recombination, and Speciation** 151
Miroslav Radman

Chapter 22 **Sexual Difficulties** 159
Howard Ochman

Chapter 23 **Unveiling *Prochlorococcus*: The Life and Times of the Ocean's Smallest Photosynthetic Cell** 165
Sallie W. Chisholm

Chapter 24 **Deciphering the Language of Diplomacy: Give and Take in the Study of the Squid-*Vibrio* Symbiosis** 173
Margaret McFall-Ngai and Ned Ruby

Chapter 25 **The Tangled Banks of Ants and Microbes** 181
Cameron R. Currie

Chapter 26 **Microbial Symbiosis and Evolution** 191
Nancy A. Moran

Chapter 27 **Coevolution of *Helicobacter pylori* and Humans** 197
Martin J. Blaser

Chapter 28 **The Library of Maynard-Smith: My Search for Meaning in the Protein Universe** 203
Frances H. Arnold

Chapter 29 **In Pursuit of Billion-Year-Old Rosetta Stones** 209
Dianne K. Newman

Chapter 30 **The Deep History of Life** 217
Andrew H. Knoll

Chapter 31 **A Glimpse into Microevolution in Nature: Adaptation and Speciation of *Bacillus simplex* from "Evolution Canyon"** 225
Johannes Sikorski

Chapter 32 On the Origin of Bacterial Pathogenic Species by Means of Natural Selection: A Tale of Coevolution 233
Philippe J. Sansonetti

Chapter 33 The Evolution of Diversity and the Emergence of Rules Governing Phenotypic Evolution 241
Paul B. Rainey

Chapter 34 The Christmas Fungus on Christmas Island 251
Anne Pringle

Chapter 35 A New Age of Naturalists 255
Rachel A. Whitaker

Chapter 36 The Ship That Led to Shape 263
Kevin D. Young

Chapter 37 Postphylogenetics 269
W. Ford Doolittle

Chapter 38 Irreducible Complexity? Not! 275
David F. Blair and Kelly T. Hughes

Chapter 39 Many Challenges to Classifying Microbial Species 281
Stephen Giovannoni

Index 287

Contributors

Frances H. Arnold
California Institute of Technology, Pasadena, CA 91125

David F. Blair
Department of Biology, University of Utah, Salt Lake City, UT 84112

Martin J. Blaser
Departments of Medicine and Microbiology, NYU School of Medicine, and Medical Service, New York Harbor Department of Veterans Affairs Medical Center, New York, NY 10016

Harald Brüssow
Nestlé Research Centre, BioAnalytical Science Department, CH-1000 Lausanne 26, Switzerland

Josep Casadesús
Departamento de Genética, Universidad de Sevilla, Seville, Spain

Sallie W. Chisholm
Massachusetts Institute of Technology, Cambridge, MA 02139

Cameron R. Currie
Department of Bacteriology, University of Wisconsin-Madison, Madison, WI 53706

Fernando de la Cruz
Departamento de Biología Molecular e Instituto de Biomedicina y Biotecnología de Cantabria (IBBTEC), Universidad de Cantabria-CSIC-IDICAN, C. Herrera Oria s/n, 39011 Santander, Spain

W. Ford Doolittle
Professor Emeritus, Department of Biochemistry and Molecular Biology, Dalhousie University, Halifax, Nova Scotia, Canada

Robert Dorit
Smith College, Northampton, MA 01063

Steven E. Finkel
University of Southern California, Los Angeles, CA 90089-2910

Kevin R. Foster
Harvard University, Cambridge, MA 02115

Stephen Giovannoni
Department of Microbiology, Oregon State University, Corvallis, OR 97331

Jessica Green
University of Oregon, Eugene, OR 97403, and Santa Fe Institute, Santa Fe, NM 87501

Eduardo A. Groisman
Yale School of Medicine, Yale Microbial Diversity Institute, New Haven, CT 06536

Diarmaid Hughes
Department of Medical Biochemistry and Microbiology (IMBIM), Box 582, Biomedical Center, Husargatan 3, 751 23 Uppsala, Sweden

Kelly T. Hughes
Department of Biology, University of Utah, Salt Lake City, UT 84112

Andrew H. Knoll
Harvard University, Cambridge, MA 02115

Roberto Kolter
Harvard Medical School, Boston, MA 02115

Richard E. Lenski
Michigan State University, East Lansing, MI 48824

Stanley Maloy
San Diego State University, San Diego, CA 92182-1010

Margaret McFall-Ngai
University of Wisconsin-Madison, Madison, WI 53706

Guido Mora
Universidad Andrés Bello, Santiago, Chile

Nancy A. Moran
Yale University, New Haven, CT 06536

Andrés Moya
Institut Cavanilles de Biodiversitat i Biologia Evolutiva, University of València; Centro Superior de Investigación en Salud Pública (CSISP) and CIBER en Epidemiología y Salud Pública, Spain

Dianne K. Newman
Howard Hughes Medical Institute, Divisions of Biology and Geological and Planetary Sciences, California Institute of Technology, Pasadena, CA 91125

Howard Ochman
Yale University, New Haven, CT 06536

Anne Pringle
Harvard University, Cambridge, MA 02115

Miroslav Radman
Faculté de Médecine, Necker, Université Paris, France

Paul B. Rainey
New Zealand Institute for Advanced Study and Allan Wilson Centre for Molecular Ecology and Evolution, Massey University, Auckland, New Zealand, and Max Planck Institute for Evolutionary Biology, Plön, Germany

Rosemary J. Redfield
University of British Columbia, Vancouver, BC, Canada

Margaret Riley
University of Massachusetts, Amherst, Amherst, MA 01003

Forest Rohwer
San Diego State University, San Diego, CA 92182

John R. Roth
Department of Microbiology, University of California, Davis, Davis, CA 95616

Ned Ruby
University of Wisconsin-Madison, Madison, WI 53706

Philippe J. Sansonetti
"Pathogénie Microbienne Moléculaire" and INSERM Unit 786, Institut Pasteur, Paris, France

Thomas M. Schmidt
Department of Microbiology and Molecular Genetics, Michigan State University, East Lansing, MI 48824

Johannes Sikorski
Leibniz Institute DSMZ-German Collection of Microorganisms and Cell Cultures, Inhoffenstraße 7B, D-38124 Braunschweig, Germany

Mitchell L. Sogin
Marine Biological Laboratory, Woods Hole, MA 02543

Evgeni Sokurenko
University of Washington, Seattle, WA 98195

Paul E. Turner
Yale University, New Haven, CT 06520

Rachel A. Whitaker
University of Illinois at Urbana-Champaign, Champaign, IL 61820

Kevin D. Young
University of Arkansas for Medical Sciences, Little Rock, AR 72205-7199

Preface

The idea of putting together a collection of essays on the general topic of microbes and evolution began to take shape during an American Academy of Microbiology colloquium in Seattle, WA, in February of 2007. Not that the topic itself was discussed at the colloquium; rather, like so many ideas, this one emerged during informal discussions over drinks and in good company. At the time there were plans afoot worldwide as to how to celebrate, in 2009, the 200th anniversary of Darwin's birth and 150th anniversary of the publication of *On the Origin of Species*. We felt that organizing a colloquium on microbial evolution at the Galapagos Islands—which indeed took place in 2009—would be a fitting tribute to the role that microbes have played in our understanding of the evolutionary process.

From the concept of the colloquium, a second idea emerged quite naturally. We should ask fellow scientists to write short, personal essays about their work. The essays would clearly manifest the investigators' enthusiasm for their work and the impact of microbes on our understanding of evolution. Moreover, because we wanted to reach a broad audience, we insisted that the essays be written in a way that would make them accessible to the general public, lacking the typical scientific jargon that sometimes makes books about science impenetrable to non-experts.

Of course, going from idea to concrete results takes time and great effort from many individuals. We were fortunate that these ideas were first voiced in the presence of Carol Colgan, then Director of the American Academy of Microbiology. Her great excitement for supporting these projects was critical to get them off the ground. Jeff Holtmeier, then Director of ASM Press, helped greatly during the early stages of planning the essay collection. Once we had the essays in hand, Michael Goldberg, ASM's Executive Director, and Patrick Lacey, Editor of ASM's magazine *Microbe,* had the idea of publishing some of the chapters in *Microbe,* a step that was very useful in bringing attention to the forthcoming collection. We are much indebted to Christine Charlip, current Director of ASM Press, and Ellie Tupper, Senior Production Editor, for their invaluable assistance in putting the finishing touches on the project.

Finally and most importantly, we are grateful to our many colleagues who agreed to step off the beaten path of scientific writing and compose these personal perspectives on their work.

Roberto Kolter and Stanley Maloy
April 2012

Microbes and Evolution: The World That Darwin Never Saw
Edited by R. Kolter and S. Maloy
©2012 ASM Press, Washington, DC
doi:10.1128/9781555818470.ch0

Darwin and Microbiology

Roberto Kolter and Stanley Maloy

Galapagos Islands, August 2009.
Charles Darwin spent a few weeks during September and October of 1835 exploring the Galapagos Islands. His observations during that time reverberate deeply in the history of science because the features of the plants and animals that Darwin saw there contributed greatly to the development of his ideas of evolution by natural selection. Of these islands he remarked in *The Voyage of the* Beagle (1939), "The natural history of this archipelago is very remarkable: it seems to be a little world within itself...." But compared to the myriad microscopic worlds present in every handful of Galapagos soil that Darwin set foot on, the archipelago would not be a "little world within itself" but a vast universe containing countless worlds of unimaginable diversity. For every grain of soil, every drop of water of our planet is rich with microbial life that lies there, waiting to be discovered. It is this vast unknown that led Edward O. Wilson, probably the most highly regarded naturalist alive today, to end his Pulitzer-winning autobiography *Naturalist* with the following lines:

> If I could do it all over again, and relive my vision in the twenty-first century, I would be a microbial ecologist. Ten billion bacteria live in

Roberto Kolter is a Professor at Harvard Medical School, Boston, MA, and *Stanley Maloy* is a Professor at San Diego State University, San Diego, CA. They jointly conceived the ideas of this collection of essays and of a trip to the Galapagos—where this Introduction was written—over a few beers many years ago.

> a gram of ordinary soil, a mere pinch held between thumb and forefinger. They represent thousands of species, almost none of which are known to science. Into that world I would go with the aid of modern microscopy and molecular analysis. I would cut my way through clonal forests sprawled across grains of sand, travel in an imagined submarine through drops of water proportionately the size of lakes, and track predators and prey in order to discover new life ways and alien food webs. All this, and I need venture no farther than ten paces outside my laboratory building. The jaguars, ants and the orchids would still occupy distant forests in all their splendor, but now they would be joined by an even stranger and vastly more complex living world virtually without end.

Indeed there are "forests of bacteria" in a grain of sand. How can one come to understand the working of these forests of bacteria that cannot be perceived through direct observation? Ecologists that venture into tropical rain forests seek understanding in two ways. First, they tabulate the species present. Second, they investigate the interactions among those species as accurately as possible. In doing so, their aim is to develop models that allow them to predict how such forests might respond to changes. Ideally, we would like to pursue the investigation of the forests of bacteria in similar ways. That is a daunting task at present because we are just beginning to get a glimpse of how extremely complex the microbial world really is! Just attempting to tabulate the number of species present in any given location reveals this remarkable complexity. Yet, describing how we learned to tabulate the number of extant bacterial species will illuminate the dramatic change in worldview regarding the planet's biodiversity that we have undergone in the last 30 years.

Already during Darwin's time naturalists were cognizant of the need to determine the relationship among species as a first step in understanding their ecology and evolution. To do so, they compared visible characteristics of each organism with those of other organisms. The classic example that comes to mind about this is the comparison of the various species of finches that Darwin noted inhabited different islands in the Galapagos archipelago. Comparing the lengths and widths of the finches' beaks, it was possible to obtain a relationship "tree" among the species—some of the species were more closely related and were assumed to have a more recent common ancestor. It is in this way that naturalists mapped the biodiversity of the planet. Ernst Haeckel, the first to coin the term "ecology" and a contemporary of Darwin, began to draw "trees of life" that illustrated the relationships among all extant living forms. These are the trees that many of us—depending on our age—received as instruction in our primary education in terms of the "kingdoms" of living forms. Perhaps many of the readers will have first been exposed to a worldview of five kingdoms: plants, animals, fungi, protists, and "monera." The relationship tree that connects these is probably familiar to many (Fig. 1).

The diagram represents not only the relationship among the five kingdoms but also the evolutionary history of the kingdoms

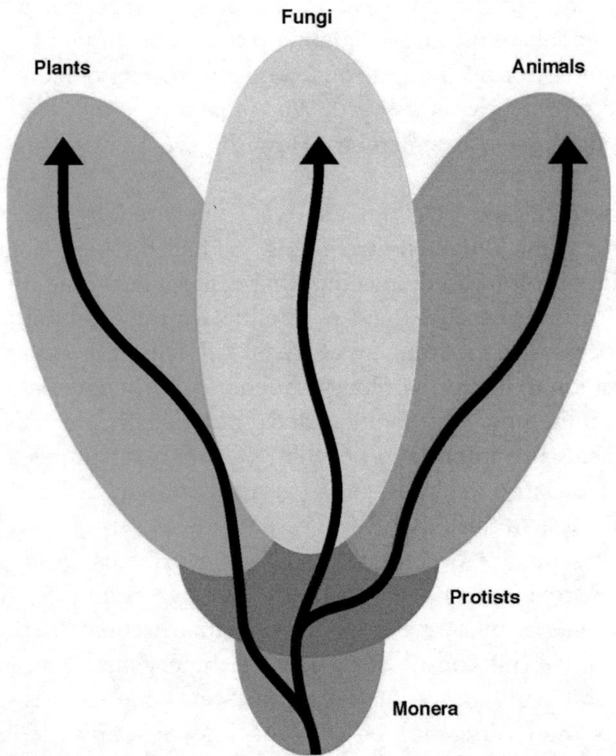

Figure 1 Relationship tree that connects the five kingdoms. doi:10.1128/9781555818470.ch0f1

and the relative diversity present in each one. In this worldview the species of plants, fungi, and animals constitute the vast majority of the planet's biodiversity and, importantly, these large visible organisms appear as having evolved from and thus being more evolved than the members of the microscopic invisible kingdoms of protists and monera. The term "monera" was indeed coined to indicate a single morphology (from *mono,* one): small. And this worldview made sense for a long time because all the comparisons were being made by comparing visible differences.

Throughout the 20th century, naturalists studying bacteria sought ways to establish relationships among different isolates beyond morphological comparisons because, in fact, the shapes of bacteria are not that many—most bacteria are either rods or spheres. Alas, most of these efforts led to frustration! But two key developments between 1977 and 1994 led to the ability to relate and identify all bacteria on the planet. Not only that, but also these developments allowed us to define the relationship, in a quantitative manner, of all living forms. And the results obtained turned our old worldview on its head. We realized then that we live on a planet dominated by microbes. How did we arrive at this understanding?

The first key development was pioneered by Carl Woese, working at the University of Illinois. He had the idea of establishing relationships based on comparing components that are present in all life forms because they are essential for the most fundamental life processes carried out by all cells. All cells carry their genetic information in the form of long sequences of different "letters" that make up the now very familiar deoxyribonucleic acid (DNA). The linear sequence information of the DNA that constitutes a gene is first made into a transient molecule known as "messenger" ribonucleic acid (mRNA). This messenger is, in turn, translated into a linear sequence of "amino acids"—the proteins. And proteins play key roles in cellular function either by serving as structural components or by carrying out most of the reactions that make all of the other cell constituents. The machinery that translates the mRNA into proteins is known as the "ribosome" and is made of proteins and a different type of RNAs, known simply as ribosomal RNAs (rRNAs). The key is that cells from *all* life forms contain ribosomes. Woese's remarkable achievement was to be the first to begin sequencing rRNA molecules from diverse forms of life. He then aligned these sequences to obtain a quantitative estimate of the sequence diversity across all forms of life. The first fact that stood out in these analyses was that, as incredible as this might sound, there are parts of the sequence of rRNAs that have been conserved throughout the entire evolution of life on Earth! This is made evident by a diagram that shows just a small segment of the over 1,500 "bases" (A, C, G, or T) of information that constitute the gene encoding one of the rRNAs (Fig. 2). The alignment shows the sequences from an animal (human), a fungus (brewer's yeast), a plant (corn), and several microbes that were all considered to be bacteria, i.e., members of the kingdom monera. In particular, note that the sequences highlighted in yellow are absolutely identical in all the organisms.

This identity of sequences permits the alignment to be made. More interesting, however, are the regions that vary. These are the ones that allow for a quantitative estimation of the differences between species. When these differences were mapped to make a "universal tree of life" there were some major surprises! The sequence diversity of the rRNA showed a dramatically different picture from that offered from the old worldview of five kingdoms. First, the tree that emerged showed that the sequence diversity

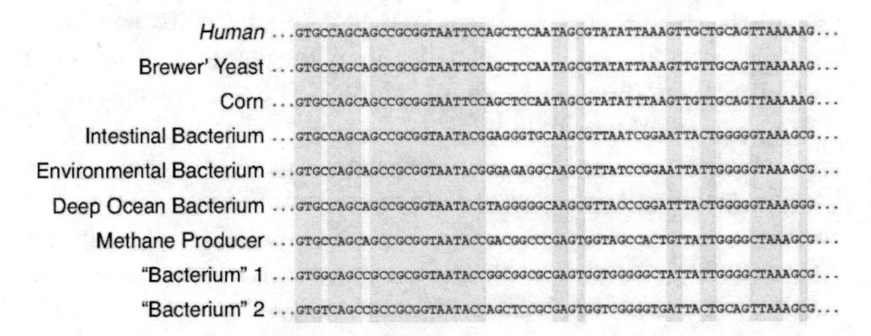

Figure 2 A small segment of the over 1,500 bases constituting the gene encoding one of the rRNAs. The alignment shows the sequences from an animal, a fungus, a plant, and several microbes that were all considered to be bacteria. Sequences highlighted in yellow are identical among all the organisms. doi:10.1128/9781555818470.ch0f2

present in all the "macroorganisms," the animals, the plants, and the fungi, was contained in a very small corner of the tree. The vast majority of the sequence diversity was represented in the microbial world. Second, the microbial world that was formerly referred to as "monera" represented two groups that were so dramatically different, it was impossible to call them all bacteria. For example, the putative bacterial sequences shown in quotes in Fig. 2 were so distantly related that Woese gave them the name "archaea" because in many cases these organisms occupied niches—such as boiling sulfur springs or extremely hot deep sea vents—that resembled conditions of primitive Earth where life is thought to have originated. This diagram shows how different our current world-view is regarding the map of biodiversity (Fig. 3).

The second key development that transformed our view of biodiversity of the planet came through the invention of a whole new way to approach the study of microbes. Since the end of the 19th century, microbes have been characterized by first isolating them and then cultivating them. This is, for example, the way most of the clinical diagnoses for many infections are carried out in microbiological laboratories. If someone is suspected of having a blood-borne infection of something like *Staphylococcus aureus*, a blood sample is sent to the clinical laboratory. There, a technician incubates that sample, giving it nutrients that allow this particular bacterium, if present, to grow. If bacterial growth is detected, the organism can be characterized. But there had always been a paradox. The number of bacteria that could be cultivated from samples was always much smaller than the number of bacterial cells that were observed in those samples. This led Norm Pace in 1994, then working at Indiana University, to attempt to isolate microbial genes directly from the environment and obtain their sequence without the need to cultivate the individual microbe. The results once again turned our worldview of biological diversity on its head. In just a few samples from extreme environments, such as those found in the hot springs of Yellowstone, he and his colleagues found more sequence diversity than what had been obtained from the entire world of cultivated microbes!

These two developments, the new quantitative and universal tree of life based on sequence comparisons and the ability to assess

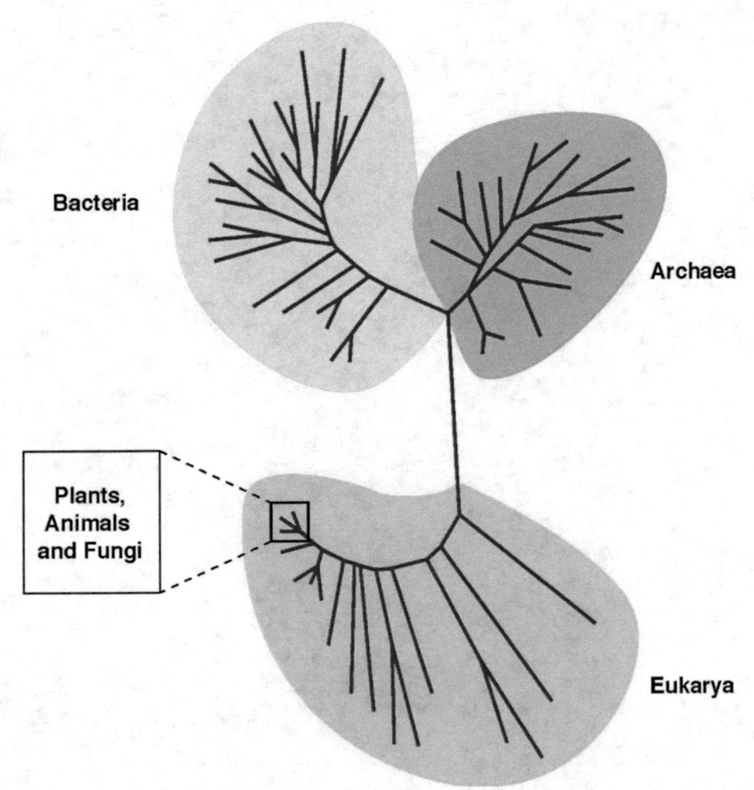

Figure 3 Map of biodiversity reflecting the current worldview. doi:10.1128/9781555818470.ch0f3

microbial diversity in samples without the need for cultivation, revolutionized microbiology in particular and biology in general. The living world ceased to be centered on macroorganisms. Instead, the fundamental role played by microbes in the natural history of our planet became cemented. These developments have led to a much greater appreciation of the role that microbes have played in the planet's evolution, and to a renewed excitement worldwide to investigate the remarkable activities that microbes carry out. In celebration of the bicentennial of Darwin's birth and the 150th anniversary of the publication of his landmark book *On the Origin of Species*, we asked 40 of our colleagues to write personal essays to communicate their work and, perhaps more interestingly, the tremendous excitement that they feel in their pursuit.

Microbes and Evolution: The World That Darwin Never Saw
Edited by R. Kolter and S. Maloy

doi:10.1128/9781555818470.ch1

1

Evolution in Action
A 50,000-Generation Salute to Charles Darwin

Richard E. Lenski

Like cuneiform on clay tablets, the history of life itself is written in minerals and in code. The minerals are fossils of long-dead organisms, while the code is the language of DNA shared by all organisms, revealing the family tree of life.

What story could be more exciting than the history of life on Earth? As an evolutionary biologist, I'm lucky to live at a time when legendary predecessors and talented contemporaries have turned the pages of this grand story by prospecting fossils and decoding genomes. Thrilling as it has been so far, I want to keep turning the pages of time. I like to watch evolution as it happens.

In 2009 the world celebrated the 150th anniversary of Charles Darwin's world-changing book, *On the Origin of Species by Means of Natural Selection*. In the closing passage of the book, Darwin emphasized that evolution is an ongoing process: "There is grandeur in this view of life, with its several powers, having been originally breathed into a few forms or into one; and that, whilst this planet has gone cycling on according to the fixed law of

Richard E. Lenski obtained his Ph.D. at the University of North Carolina, Chapel Hill, and did postdoctoral research at the University of Massachusetts, Amherst. He was at the University of California, Irvine, before moving to Michigan State University in 1991. He enjoys old books, new ideas, and especially his granddaughter Ilana.

gravity, from so simple a beginning endless forms most beautiful and most wonderful have been, and are being, evolved."

But can anyone really observe evolution in action? Though Darwin realized that evolution was ongoing, he emphasized its gradual nature: "It may be said that natural selection is daily and hourly scrutinising, throughout the world, every variation, even the slightest; rejecting that which is bad, preserving and adding up all that is good… We see nothing of these slow changes in progress, until the hand of time has marked the long lapse of ages, and then so imperfect is our view into long past geological ages, that we only see that the forms of life are now different from what they formerly were." Darwin also recognized the power of humans to exert selection and cause evolution—he began *Origin* by explaining how humans had domesticated animals and plants by breeding those with desirable features, in order to introduce his idea of natural selection—but even these changes he thought would be imperceptible "...unless actual measurements or careful drawings of the breeds in question had been made long ago, which might serve for comparison."

Even so, one of his readers pressed ahead with the idea of watching evolution in action. The Rev. William Dallinger (1839–1909) was not only a Methodist minister, he was also skilled in the methods of microbiology. For several years he grew protozoa in an incubator, gradually raising their temperature. The organisms he used to start the experiment struggled even at 73°F, while those at the end tolerated 158°F but were unable to grow at the initial temperature of 60°F.

Dallinger showed that it was possible to observe the process of evolution over a human timescale by studying fast-reproducing microbes. Others also saw the possibility of studying evolution in action. In 1892, Henri de Varigny published a book called *Experimental Evolution* in which he proposed long-term experiments that would outlast the lifetimes of the participating scientists. Beginning early in the 1900s, fruit flies in the genus *Drosophila* became widely used for genetics research, and experiments were performed that demonstrated the effects of natural selection and random genetic drift.

Bacteria had to wait in the wings for many years before they could star in evolution experiments. While the science of genetics took hold with the rediscovery of Gregor Mendel's experiments on pea plants, the experts were baffled by the question of heredity in bacteria. Microbiologists saw that bacteria could adapt to challenges, but they couldn't tell whether spontaneous mutants had been selected or, alternatively, whether the challenge had induced the cells to change themselves. In 1934, a microbiologist, I. M. Lewis, wrote that "The subject of bacterial variation and heredity has reached an almost hopeless state of confusion... There are many advocates of the Lamarckian mode of bacterial inheritance, while others hold to the view that it is essentially Darwinian." In 1942, Julian Huxley wrote a book entitled *Evolution: The Modern Synthesis* that excluded bacteria from that synthesis on the grounds that "[t]hey have no genes in the sense of accurately quantized portions of hereditary substance...."

This confusion cleared the next year with the publication of what is, to me, the single greatest experiment in the history of biology. Working as a team, a biologist, Salvador Luria, and a physicist-turned-biologist, Max Delbrück, employed subtle reasoning and an elegant design to demonstrate that certain mutations in *Escherichia coli* occurred *before* the selective challenge was imposed and hence could not be changes induced *by* the challenge. In other words, mutations are random events that occur whether or not they will prove useful, while selection provides the direction in evolution by retaining mutations that are advantageous to their bearers and discarding others that are harmful.

Luria and Delbrück's paper launched a tidal wave of research that led to the discovery of DNA as the hereditary material and to cracking the genetic code, among other achievements. But it had little immediate impact on evolutionary research. The new molecular biologists pursued their reductionist methods, while evolutionary biologists, grounded in natural history, didn't want to study things they couldn't even see. These naturalists preferred beautiful butterflies and even homely fruit flies to *E. coli* organisms that, after all, come from a rather uninviting habitat. It was also difficult to tell bacterial strains and species apart, and many evolutionary biologists were focused on using patterns of similarities and differences to unravel the relationships among organisms.

But eventually the field of microbial evolution awakened, and for several reasons. In the 1970s, Carl Woese used differences in DNA sequences to study the evolutionary relationships among bacteria and other microbes, revealing extraordinary diversity beneath their outwardly simple appearances. Meanwhile, pathogenic bacteria that had been successfully treated with antibiotics often evolved resistance to those drugs, while other microbes emerged as pathogens, sometimes by acquiring new capabilities. In time, a few visionaries realized that microorganisms could be used—just as the Rev. Dallinger had foreseen a century earlier—in experiments to test Darwin's ideas and, more generally, evolutionary theory as it had developed over that century.

I was drawn into this field as a postdoc in the early 1980s. I had done my doctoral research in zoology, studying insects in the mountains of North Carolina. Despite the pleasures of working outdoors, data collection was slow, heavy rains drowned my beetles in their pitfall traps, and it was difficult to imagine feasible experiments that would really test the scientific ideas that most excited me. As I pondered future directions, I remembered the beautiful experiment by Luria and Delbrück that I had encountered as an undergraduate. I recalled not only its elegance but also the insight it provided into the tension between randomness and direction in evolution. Evolution is like a game that combines luck and skill, and perhaps bacteria could teach me some interesting new games.

Games that involve both luck and skill rarely play out the same way twice, and that uncertainty is part of their fascination. Using bacteria, I could watch replicate populations—all starting with the same ancestral strain, and all living in identical environments—to see just how similarly or differently they would evolve. So in 1988, I started an experiment with 12 populations of *E. coli* that I intended to keep going for at least 2,000 generations, maybe longer.

Today, the bacteria have been evolving in and adapting to their separate little worlds for over 50,000 generations. Each population lives in a flask containing 10 milliliters of a solution containing the sugar glucose as the limiting resource. Each day, including weekends and holidays, someone in my group withdraws one-tenth of a milliliter from a culture and transfers that into 9.9 milliliters of fresh medium. The bacteria grow until the glucose is depleted, and then they sit there until the same process is repeated the next day. Bacteria grow by binary fission—that is, a cell grows in size before dividing into two daughter cells—so that the 100-fold dilution and regrowth allow almost seven doublings, or generations, per day. Although this might sound like an easy existence for the bacteria, there is fierce competition to get the glucose and grow faster than anyone else. And every day, 99% of each population is consigned to oblivion by the random draw of a drop into a pipette that determines which lucky cells will continue this small, but great, struggle for existence.

However, survival and oblivion aren't the only two fates for these bacteria. A third possibility—suspended animation—is one of the most important features of this experiment. Every so often, instead of discarding the leftover bacteria from the previous day, we add a cryoprotectant and store the cells in a freezer. The result is an extraordinary "fossil record" where we can revive the bacteria and compare living cells from different generations. We can even compete bacteria against their own ancestors from thousands of generations earlier. These competitions allow us to measure the improved adaptation of the bacteria to their flask world that has resulted from natural selection—the process that Darwin realized would give rise to organisms that were well fit to their environment. Imagine Neanderthals brought back to live among us. How would they fare at chess or football? How far have we come in the nature of our genes and the nurture of our culture? When I switched my research to bacteria, their speedy reproduction was an obvious attraction for watching evolution in action. But I've come to realize that the ability to freeze and revivify the bacteria—to study true living fossils—is just as important to my research.

When I began this experiment, I thought big differences among the 12 lines would soon be apparent. The random occurrence of mutations meant that some populations would get lucky by generating a beneficial mutation (and one that survived the daily dilutions) sooner than others. And just as in a game, different early moves—mutations—might open some doors while closing others. Some populations might get stuck with beneficial mutations that ultimately led nowhere, while others would follow paths that had long-term potential.

To my surprise, evolution was pretty repeatable. All 12 populations improved quickly early on and then more slowly as the generations ticked by. Despite substantial fitness gains compared to the common ancestor, the performance of the evolved lines relative to each other hardly diverged. As we looked for other changes—and the "we" grew as outstanding students and collaborators put their brains and hands to work on this experiment—the generations flew by. We observed changes in the size and shape of the bacterial cells, in their food preferences, and in their genes. Although the lineages certainly diverged in many details, I was struck by the parallel trajectories of their evolution, with similar changes in so many phenotypic traits and even gene sequences that we examined.

At the same time, we pursued other evolution experiments with *E. coli*. We varied their diets and the temperatures of their flasks. We allowed some to have sex by introducing special genes that let cells conjugate and recombine their genomes. We branched out and studied another bacterial species, *Myxococcus xanthus*, that has fascinating behaviors in which cells cooperate to form multicellular fruiting bodies as well as to hunt down and consume our beloved *E. coli*. Social norms evolved, including the appearance of cheaters that benefit from belonging to a group but harm the group's overall performance. We saw the evolution of predators that swarmed outward and found patches of their prey more quickly than their ancestors.

But back to the main story. The generations marched by, thousands upon thousands. The rate of improvement progressively slowed as the *E. coli* became better adapted to their new world. My colleagues began to track down the mutations responsible for that adaptation. Those analyses required skills I lacked, as new technologies emerged that could be applied to our experiment. As for myself, I branched into a new line of research.

Perhaps it was a mid-life crisis: my bacteria were slowing down, and I was looking for some new action. So I had an affair—one that continues today, though with slightly less feverish intensity—with some artificial creatures. Avidians are computer programs that copy their own genomes, and they live in a virtual world that exists inside a computer. But their replication is imperfect, so Avidians sometimes mutate. While most mutations are deleterious, some provide an advantage that allows a mutant to obtain resources and replicate faster than its competitors in that virtual world. Because they start in a primitive state—the ancestral Avidian was written by a computer scientist who endowed it with the capacity to replicate, but gave it no other function—we could watch not just subtle improvements but also the emergence of brand-new capabilities as they evolved a rich computational metabolism. However, it seems that the *E. coli* became jealous of the attention I lavished on the Avidians, because one lineage decided to show me that it could do something new, too.

The long-term experiment was designed to be a simple one in which time and the bacteria would do the work of evolution. In that spirit, I had chosen to propagate the lines in an environment with just one sugar, a constant temperature, no predators—in essence, just about as simple as I could make it. So while there were many opportunities for improvement, it couldn't be expected that the bacteria would learn some entirely new trick. What would be the point? The bacteria were becoming specialists—*Escherichia erlenmeyeri*, I sometimes call them—exquisitely adapted to a simple life in an Erlenmeyer flask, as opposed to the complicated world of the colon they had left behind.

However, I had left them an opening. Although glucose is the only sugar in their environment, another source of energy, a compound called citrate, was also there all along as part of an old microbiological recipe. One of the defining features of *E. coli* as a species is that it can't grow on citrate because it's unable to transport citrate into the cell. For 15 years, billions of mutants were tested in every population, but none produced a cell that could exploit this opening. It was as though the bacteria ate dinner and went straight to bed, without realizing a dessert was there waiting for them.

But in 2003, a mutant tasted the forbidden fruit. And it was good, very good. The descendants of that mutant rose to dominance owing to their access to that second course. At first, we thought this flask had been contaminated by some other species that consumed citrate. However, DNA tests showed that the citrate-eating cells were descendants of the *E. coli* ancestor used to start the experiment.

The citrate eaters still eat glucose, but they aren't quite as successful at competing for that sugar as they were before. As a consequence of that tradeoff, their cousins persist as glucose specialists. So the bacteria in this simple flask world have split into two lineages that coexist by exploiting their common environment in different ways. And one of the lineages makes its living by doing something brand-new, something that its ancestor could not do.

That sounds a lot like the origin of species to me. What do you think?

Happy anniversary, Mr. Darwin! Over 150 years ago you revealed your ideas to the world, and evolution continues to fascinate. And remarkably, we can now observe "these slow changes in progress" even before "the hand of time has marked the long lapse of ages."

Microbes and Evolution: The World That Darwin Never Saw
Edited by R. Kolter and S. Maloy

doi:10.1128/9781555818470.ch2

2

Minimal Genomes and Reducible Complexity

Andrés Moya

Minimal Cells: from Aristotle to Plato

A review of Aristotelian reality, represented by the diversity of bacterial cells with small and reduced genomes which evolved naturally, can help delineate the Platonic idea of a hypothetical minimal cell. The quest for the minimal cell could be defined as the search for those features that are necessary and sufficient for life. Nowadays, complete genome sequences are known for many species. Furthermore, the development of in vitro as well as in silico analytical tools enables the study of molecules that make up the intricate machinery of cells, thus bringing us nearer a breakthrough in our understanding of biological systems. Besides such advances, the evolutionary perspective of cells provides the framework for a better understanding of the properties of minimal living systems.

The approach takes an existing organism and aims to simplify its genome, thereby reaching a minimal (or at least reduced) gene set. The underlying rationale is to determine the minimal requirements for life in a particular environment. Essential genes can be

Andrés Moya holds a Ph.D. in Biology and Philosophy and is Professor of Genetics at the University of Valencia. He was a postdoctoral fellow and guest professor at the University of California, at Davis and Irvine, respectively. Like a true Mediterranean he enjoys good food, and above all he is a voracious reader of philosophy essays.

identified via in silico comparative genome analysis or through deletion mutant experiments. The difference between computational and experimental approaches is that the former identifies a set of essential genes shared by different taxa, whereas the latter searches for individual genes that are essential for growth in a single organism and in the conditions under study. Thus, a theoretical minimal genome can be defined as that comprising the identified set of essential genes. Systematic attempts to identify essential genes through gene inactivation experiments have been carried out in *Escherichia coli* (620 out of 3,746), *Bacillus subtilis* (271 out of ~4,100), and *Mycoplasma genitalium* (382 out of 482). The recent synthesis of the complete genome sequence of an *M. genitalium* strain is also an effort towards engineering simpler biological systems, taking present-day cells as a starting point.

Naturally Evolved Reduced Genomes

The study of minimal cells can benefit enormously from the study of present-day organisms with small genomes by showing how relatively simple biological systems have evolved and currently operate. Thus, cells and reduced genomes of endosymbionts, parasites, and free-living organisms are examples of naturally evolved minimal gene sets. Given that small genomes retain the basic functions to sustain a cell, comparative analysis of their diverse biology may shed light on the properties of a hypothetical minimal cell.

The definition of a minimal cell is context dependent. We must make a distinction between minimal cells growing in chemically complex environments (heterotrophic) and those demanding only minimal components from their environment (autotrophic) to construct their molecules. Not unexpectedly, the fewer genes in a reduced genome, the greater chemical complexity required for the cell to grow. Obligate intracellular host-associated prokaryotes have the smallest genomes, with endosymbionts (prokaryote symbionts living inside a eukaryotic cell) harboring the smallest genomes.

Several species of free-living prokaryotes belonging to different orders have evolved small genomes with a similar number of genes, indicating that cells require around 1,400 to 1,500 genes to sustain a free-living lifestyle under current natural conditions. These observations, although limited by the data available, are nevertheless supported by the diversity of lineages, as well as the different nutritional strategies and ecological niches they represent. Currently, the free-living organism with the sequenced genome harboring the fewest genes is an uncultured but abundant ocean member of the *Betaproteobacteria*, HTCC2181, with 1,377 genes. This small genome is closely followed by others, like those of the cosmopolitan oceanic bacterium "*Candidatus* Pelagibacter ubique" (1,394 genes and one of the most successful clades of bacteria), the dehalorespirant *Dehalococcoides ethenogenes* (1,436 genes), and the hyperthermophilic crenarchaeon *Ignicoccus hospitalis* (1,494 genes). The diversity of lineages, nutritional strategies, and ecological niches occupied by these free-living organisms with small genomes is noteworthy. For instance, HTCC2181 is an obligate methylotroph marine bacterium that specializes in using one-carbon compounds like methanol and formaldehyde as sources of carbon and energy. "*Candidatus* P. ubique" is a heterotroph that grows by assimilating organic compounds dissolved in the ocean, and can produce energy by a light-driven proteorhodopsin proton pump or by respiration. *D. ethenogenes* is a strictly anaerobic bacterium that can derive all its energy from the reduction of vinyl chloride to ethene. The smallest photosynthetic cells, represented by *Prochlorococcus marinus*, have slightly larger genomes (1,660 kb and 1,765 genes). The hyperthermophilic bacterium *Aquifex aeolicus* (1,613 genes), a chemolithoautotroph, can grow on a very simple chemical menu of hydrogen, oxygen, carbon dioxide, and mineral salts. Among archaean prokaryotes, *I. hospitalis* is another chemolithoautotroph that couples CO_2 fixation with sulfur respiration using molecular hydrogen in high-temperature hydrothermal vents. *Methanococcus aeolicus* (1,555 genes) is a CO_2-reducing methanogenic marine member of the *Euryarchaeota*. Another member of the *Euryarchaeota*, the strictly aerobic extremophile *Picrophilus torridus* (1,605 genes), is a heterotroph that can use propionate as well as many types of sugar as an energy source and can grow at pH 0 and up to 65ºC, thus representing one of the most thermoacidophilic organisms known. Last, but not least, is the aerobic hyperthermophilic korarchaeon "*Candidatus* Korarchaeum cryptofilum" (1,660 genes), which grows heterotrophically using a variety of peptide and amino acid degradation pathways.

Genomes with fewer genes than the smallest free-living prokaryote belong to parasitic or endosymbiotic organisms and are found in 15 different orders among currently sequenced genomes. Having evolved from free-living relatives by a process of massive genome reduction, they have traditionally been studied as natural minimal cell models. Perhaps one of the most striking findings is that these reduced genomes lack genes previously deemed essential for life. For example, a strain of "*Candidatus* Phytoplasma asteris" and *Buchnera aphidicola* BCc (primary endosymbiont of the cedar aphid *Cinara cedri*) lack the genes coding for the ATP-synthase subunits. *Ureaplasma urealyticum* lacks the heat shock protein/chaperonins GroEL and GroES and the cell division protein FtsZ. Examples of simpler biochemical pathways are also found in *U. urealyticum*. This bacterium generates 95% of its ATP through the hydrolysis of urea by urease. In terms of the number of genes involved, ATP production by urea hydrolysis is a simpler process than carbohydrate metabolism. Comparing the genome of *Mycoplasma* mycoides with the minimal gene set (254 genes) from *M. genitalium* and *Haemophilus influenzae* shows that 11 out of 254 genes are absent in this species. Furthermore, several genes shown to be essential in *E. coli* are absent in the related gammaproteobacterium symbiont of the clam *Calyptogena okutanii*, including genes for cell division (*ftsZ* and related genes), among others.

The Flagellum of a Bacterial Endosymbiont: A Case against Intelligent Design

As already stated, minimal cells, and in particular their reduced genomes, have evolved from ancestors possessing a larger repertoire of genes. The type of genes retained as well as their putative functions may be clues to understanding the evolution of complex genetic features in their ancestors. This is the case of the flagellar apparatus in *Buchnera aphidicola*, the primary endosymbiont of aphids.

Aphids feed on sap, which is rich in sugars but lacks other nutrients (for example, it is deficient in essential amino acids and vitamins). Bacteria of the genus *Buchnera* live in association with aphids, and in exchange for a protected environment and guaranteed sugar supply, these microorganisms synthesize essential amino acids and vitamins, both for themselves and for their insect host. Such a protected and nutrient-rich environment has led *Buchnera* organisms to lose many of their genes (the smallest known genome has only 422 genes). Figure 1 shows the genes coding for the flagellar apparatus, which are still present in the smallest *Buchnera* organism (Fig. 1b), compared with the genes for the homologous flagellum in its free-living relative *E. coli* (Fig. 1a). This particular *Buchnera* organism (*B. aphidicola* BCc) is an endosymbiont in an advanced stage of genome reduction and with highly simplified cell envelopes, also lacking a typical type III secretion system. Notwithstanding, it retains surface mechanisms for exporting proteins, such as the simplified flagellar apparatus. These structures are quite abundant on the *Buchnera* surface even though the bacteria are nonmotile; furthermore, different *Buchnera* organisms have retained different parts of the flagellum but all have conserved those elements homologous to the type III secretion system. Consequently, these structures are thought to be involved in invading new bacteriocytes (specialized host cells that lodge endosymbionts), ovaries, and embryos to ensure transmission to the host's offspring. *B. aphidicola* BCc has lost all the genes involved in synthesis of the flagellar filament, hook, and stator, none of which are necessary for this nonmotile bacterium. Nevertheless, it has retained the complete set of genes that are homologous to type III virulence secretion systems: those for FliF (the membrane-embedded MS ring); FlhA, FlhB, FliOP, FliQ, and FliR (integral membrane export components inside the MS ring); FliI and FliH (ATPase and regulator); and FliG and FliN (two out of three components of the C ring, the switch complex).

The P ring (FlgI) and L ring (FlgH) are also present. There is evidence suggesting that homology exists for FlgI (with a group of proteins called secretins) and the lipoprotein FlgH (with lipoprotein chaperones of secretins). FlgH is thought to be homologous to

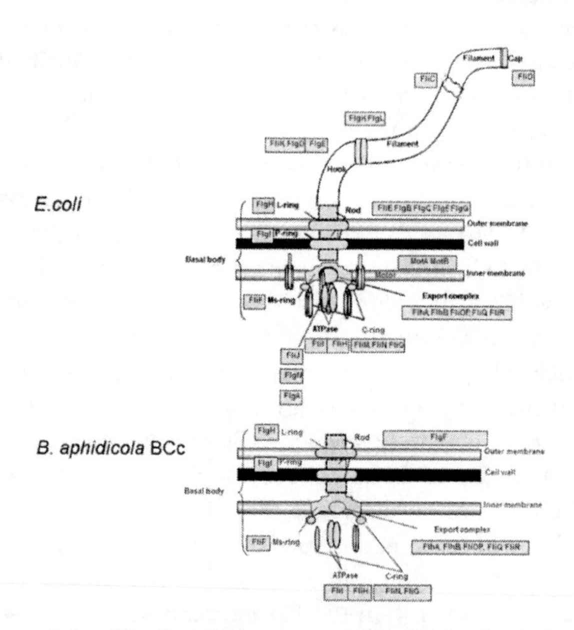

Figure 1 The flagellum, an example of tinkering in bacterial evolution. (A) *Escherichia coli* flagellum. Protein and structural components are indicated: filament, hook-filament junction, hook, and basal body. The last is formed by a set of rings anchoring the inner and outer membranes and the cell wall, the rod, the stator, the rotor, and the export apparatus (including one ATPase). (B) The reduced flagellar apparatus in *Buchnera aphidicola* BCc, from the cedar aphid. Elements involved in the formation of the flagellum filament and hook and the stator/motor have been lost. None of them are necessary for a nonmotile bacterium. Many elements involved in the export apparatus anchorage to the cell envelope have also been lost, consistent with the absence of a well-structured outer membrane and cell wall. All the retained proteins have a homolog of the type III secretion system, supporting the hypothesis that this structure is used by the bacterium as a protein export system to facilitate the invasion of new host cells. doi:10.1128/9781555818470ch2f1

the *Salmonella* type III secretion system protein InvH, a lipoprotein required for insertion of the InvG secretin in the outer membrane.

Only one of four proteins for rod formation (FlgF) is preserved. These components are highly conserved across all known bacterial flagella, but their dispensability for building a basic filament is shown by type III virulence systems, in which no rod homologs have yet been discovered even though the pilus protein shares similarities with axial proteins. The four rod proteins (FlgBCFG) seem likely to be duplicates of an ancestral rod protein. Therefore, possibly just one of them is able to form a central channel to export targeted proteins across the outer membrane. However, *Buchnera* has lost all necessary enzymes to synthesize an outer membrane, and therefore, the rod protein FlgF might only be necessary to keep the P and L rings in the proper position. Finally, the membrane-associated protein FliJ, which bears similarities with the type III cytoplasmic chaperone family, is missing.

Proponents of intelligent design have argued that complex apparatus like the flagellum are examples of irreducible complexity. They state that all the elements composing a given complex structure, like the flagellum, must always be present for it to work. However, the reduction of the flagellar apparatus in *Buchnera* is a wonderful lesson in evolutionary tinkering that shows us its convoluted history in two main respects. First, certain components of the retained flagellar genes serve not for bacterial mobility but, rather, to export proteins that could eventually become involved in infecting new surrounding host cells, ovaries, or embryos, thereby enabling *Buchnera* to be vertically transmitted to its host's offspring. Second, and probably more importantly, *Buchnera* has recovered functions that were present in the ancestors of current free-living relatives, which possess a complete flagellum (Fig. 1a). Thus, the particular genes now retained for *Buchnera*'s reduced flagellum would have evolved together with other genes involved in different functions up until the formation of a complete flagellum. If this is the case, the flagellum would be an example of reducible complexity.

Microbes and Evolution: The World That Darwin Never Saw
Edited by R. Kolter and S. Maloy

doi:10.1128/9781555818470.ch3

3

Lady Lumps's Mouthguard

Jessica Green

My name is Thumper Biscuit. I am a blocker on the Flat Track Furies, a roller derby team in Eugene, OR. I aspire to be a jammer. This will require that I overcome the instinct to cower before being plowed over by a fast-moving woman on wheels. There are up to 10 women on wheels colliding at any given time on the derby track. Survival requires being acutely aware of the spatial coordinates of all players at all times. Fortunately, I am a biogeographer by day—I study the spatial distribution of biodiversity (the variety of life). For several years I have been consumed by the biogeography of microbial life. The more I learn about microorganisms, the more I find myself imagining where they are around me and what they are doing. Even while skating I think about microorganisms. I know I should be listening to our coach, One Eyed Jack, and concentrating on new team strategies, like the "bullshit," which requires finessing a double-whip around the opposing team blockers. Instead I daydream about the variety of microbes dispersing through the air and hitting my face as I skate. I wonder how the microbes down by my wheels are evolutionarily related to those above my helmet, and which are better adapted to Lady Lumps's mouthguard than mine.

Jessica Green was academically trained at UCLA, UC Berkeley, and UC Davis. She was founding faculty at UC Merced, and moved to the University of Oregon in 2007. She became a jammer on the Flat Track Furies for two years before retiring to play chess and cook with her sons Mauro and Max.

If Charles Darwin had known that microbes comprised the vast majority of life on Earth, he probably would have asked similar questions (although not while rollerskating). His research on the spatial distribution of macroorganisms (plants and animals) formed an integral part of his arguments for natural selection and played a foundational role in the burgeoning field of biogeography. Darwin recognized three biogeographic patterns that he described in terms of "great facts" in *On the Origin of Species*. In summary, these are: (i) environmental conditions alone cannot account for the dissimilarity of flora and fauna among geographically distinct regions; (ii) barriers to dispersal significantly contribute to these differences; and (iii) although the spatial variability in community composition within regions is substantial, these communities remain evolutionarily related. One hundred fifty years have passed, and scientists are just now beginning to appreciate the relevance of these "facts" to microbial life.

The prevailing view of Darwin and his contemporaries was that microorganisms are dispersed globally and able to proliferate in any habitat with suitable environmental conditions. *On the Origin of Species* proposes that "the lower any group of organisms is, the more widely it is apt to range," a notion that was crystallized by Lourens Baas-Becking in the 1930s by his widely referenced quote "everything is everywhere, but the environment selects." The small size and high abundance of microbes (as well as other aspects of their biology) were thought to increase the rate and geographic distance of dispersal to levels such that dispersal limitation is nonexistent, resulting in fundamentally different biogeographic patterns from those of plants and animals. Testing the universality of these assumptions was difficult (if not impossible) because prokaryotic and many eukaryotic microorganisms cannot be identified morphologically and, until recently, could only be identified using culture-based methods in the laboratory (for example, by growing them on an agar plate). We now know that only a small fraction of microbial life—some say less than 1%—can be detected using culturing techniques. This limitation undoubtedly skewed the world's understanding of microbial diversity and geography. Consider the potential shift in Darwin's ideas if he were confined to "seeing" 1% of the macroscopic species in the landscapes he traversed. Would he have arrived at the theory of evolution by natural selection?

Recent advances in our ability to characterize microbial diversity in natural environments have prompted a new era of exploration, one that has radically changed thinking in ecology and evolution. Scientists have moved beyond the limitations of culture-based techniques and are using genes as a yardstick to quantify the evolutionary relatedness among Earth's organisms (the "Tree of Life"). The gene-based Tree of Life significantly differs from the five-kingdom model I was taught in high school, which was composed of animals, plants, fungi, protists, and monera (prokaryotes). We now group life into three major branches: *Eukarya* (including plants, animals, and microscopic eukaryotes), *Bacteria,* and *Archaea* (both comprising microscopic prokaryotes). This molecular perspective shows that most of the genetic diversity on Earth is made up of microbes—organisms invisible to the naked eye.

Biologists spanning the globe have enthusiastically embraced molecular tools to explore microbial diversity in every imaginable habitat: oceans, lakes, dirt, air, leaves, stomachs, and more. I have joined the hunt, using gene probes in attempts to understand if microbes have biogeographic patterns similar to those commonly observed for plants and animals. One of the most general, best-documented patterns in nature is the species-area relationship, which shows that species number tends to increase with increasing area. Advocates of the "everything is everywhere" hypothesis suggested that unique aspects of microbial biology might prevent microbes from exhibiting species-area relationships. As part of a team of scientists in the rolling red sand dunes of Australia, I found that microbes are spatially distributed across the landscape in a manner similar to that of macroorganisms and that they also exhibit a species-area relationship.

Another striking characteristic of life on Earth—recognized since at least the early 1800s—is the increase of species richness from the poles to the equator. Some argued that the latitudinal diversity gradient should be reduced or absent for unicellular organisms, due to their high abundance and dispersal potential. Collaborating with marine biologists who had collected oceanic samples throughout the world, we found that bacteria showed geographic patterns of diversity similar to those reported for larger organisms, with bacterial species richness twice as large in the tropics relative to the poles. These studies and many others provide evidence that microbial biogeography is not fundamentally different from that of other life forms. Rather, microbial diversity shifts across the surface of the Earth as a consequence of the three forces highlighted by Darwin when considering plants and animals: the environment, dispersal, and diversification.

That microbes continue to be at the forefront of contemporary biogeography research is not surprising considering their extraordinary abundance and diversity. Researchers have estimated that there are more than 6 billion microbes in the human mouth, comprising more than 600 species. On Earth, the number of individual microbes is thought to be as high as 10^{30}, with a speculated 10 million to 1 billion species. Scientists have gone beyond cataloguing microbial diversity to explore how this diversity links to the function and "health" of ecosystems. For me, the thrill of studying microbial communities (what got me hooked) is the promise of conceptualizing biodiversity in novel ways.

The first novelty posed by working with microbes, given my background in community ecology, was considering biogeography in an evolutionary context. Thinking in this context is not new for many biologists, nor is it limited to the study of microbes. Darwin was clearly thinking about the spatial scaling of evolutionary diversity when he wrote on his observations of birds and rodents, "the naturalist in travelling from north to south never fails to be struck by the manner in which successive groups of beings, specifically distinct, yet clearly related, replace each other." Microbes focused my attention on evolutionary patterns, because rather than macroorganism species lists, my samples comprised a list of genes. Translating a list of genes into a list of species, or taxonomic units, requires going through a process that simultaneously generates information on the evolutionary relatedness among individuals. This makes it both attractive and tractable to quantify metrics of evolutionary diversity, for example, the phylogenetic diversity within a region (measured as the length of evolutionary pathways that connect a given set of taxa) and the phylogenetic similarity between regions (measured as the length of evolutionary pathways shared between taxa in two regions). Together, these two diversity metrics offer important clues about the relative importance of key ecological and evolutionary processes—such as environmental filtering, immigration, and diversification—to the distribution and abundance of taxa in space and time. Phylogenetic information garnered from microbial communities also presents an unprecedented opportunity to answer conservation questions about the magnitude of evolutionary diversity on Earth or the pace of biodiversity loss in response to global change.

The second novelty has been rethinking existing theories of biodiversity and biogeography. Despite the recognized importance of evolutionary diversity, little is known about how it changes from the scale of a seawater droplet to the Atlantic Ocean or between disparate airborne particles in a room. This is partially due to our inability to sample exhaustively the biotic world at ever-increasing scales. Limitations on time, resources, and sequencing efficiency have constrained the spatial extent of evolutionary diversity studies. Coupled with these limitations is the current lack of ecological and evolutionary theory to predict spatial patterns of evolutionary diversity.

My laboratory is striving to overcome these limitations by developing a spatial theory of community assembly that quantifies the evolutionary relatedness among individuals (or genomes or genes) within and among sample locations across the landscape. This new framework combines four crucial ingredients: spatial dispersal, by which individuals disperse seeds or move across a spatial landscape; evolutionary diversification, through which genetic novelty is introduced to a community; individual-based processes, so that the discrete nature of individuals is accounted for; and demographic stochasticity, the random nature of birth, death, and dispersal processes. If successful, the outcome will be spatially explicit theoretical predictions, including how evolutionary diversity increases with spatial scale, and the evolutionary similarity between two spatially separated regions.

One of the Challenge Bouts at Rollercon 2009 in Vegas was *Amazon* versus *Shortbus* (players taller than 5′ 10″ versus those smaller than 4′ 11″). Vexine, one of our best blockers, wanted to gather data on the spatial dynamics of the players to assess which body size is better fit for the derby environment. A compelling question, but I'd rather gather data on the spatial dynamics of the invisible players on the track. I have already started collecting skin samples. Next year: mouthguards.

Microbes and Evolution: The World That Darwin Never Saw
Edited by R. Kolter and S. Maloy

doi:10.1128/9781555818470.ch4

4

Trying To Make Sense of the Microbial Census

Mitchell L. Sogin

What do microbes and stars have in common? Their numbers are staggering, and most can only be seen through high-powered optical devices. Based upon observations with powerful telescopes, astronomers teach us that the Milky Way has more than 10^{12} stars and that 10^{12} similar-size galaxies account for the more than 10^{24} stars in the universe—roughly equivalent to the number of grains of sand on Earth. Yet nearly every one of those sand grains harbors hundreds of thousands, if not millions, of microbial cells visible through high-resolution scanning electron microscopes. Thus, the number of microbes on Earth eclipses the number of stars in the universe by many orders of magnitude. And the collective biomass of single-cell organisms on Earth outweighs all of the plants and animals combined. Single-cell organisms occupy every imaginable niche ranging from the deep subsurface to the microbiomes of multicellular organisms. From the time of their origins, microscopic factories have served as essential catalysts for all of the chemical reactions within biogeochemical cycles that shape planetary change and habitability. They regulate the composition of the atmosphere,

After earning a Ph.D. at the University of Illinois, *Mitchell Sogin* did postdoctoral training at the National Jewish Hospital in Denver, joined the faculty of The University of Colorado Health Sciences Center, and became a professional skier. In 1966 he founded the Bay Paul Center at the Marine Biological Laboratory, Woods Hole, where he spends his weekends sailing on *Origins*.

influence climates, recycle nutrients, and decompose pollutants. Without microbes, multicellular life on Earth would not have evolved and biology as we know it would not be sustainable.

Given the massive number of microbes with seemingly unlimited metabolic diversity, the accumulation of mutations during the past 3.5 billion years could have led to enormous numbers of distinct microbial populations that exhibit high levels of genetic diversity and phenotypic variation. Yet, practitioners of biodiversity (a form of political ecology) often treat the microbial world as an afterthought, with only 5,000 named taxa. Nearly 100 years after the discovery of the microbial world, scientists could only identify and classify microbes by morphology, cell staining characteristics, physiological properties, and the chemical reactions they perform. The limited morphological complexity discernible through light microscopy and the requirement to grow microbes in the laboratory for identifying phenotypic features that can serve as taxonomic determinants constrained our ability to observe and differentiate between kinds of microbes. The lack of agreement about which phenotypic features were most valuable for defining microbial systematics led to conflicting views of their natural relationships and gross underestimates of diversity.

By the 1960s, many microbiologists had given up on the goal of achieving a coherent "tree of life" for single-cell organisms. It took a physicist-turned-microbiologist at the University of Illinois at Urbana to arrive at a solution. Carl Woese argued that molecular data could provide a practical metric for assessing evolutionary relationships between cultivated microorganisms in the context of 3.5 billion to 3.8 billion years of evolutionary history. He reasoned that the RNA component of the cell's protein synthesizing machinery evolved very slowly because it interacted with all other proteins in the cell either during their synthesis or as part of the ribosome, the nucleic acid-protein complex that assembles amino acids into proteins by a nucleic acid templating process. Because of its slow pace of evolution, similarities between ribosomal RNA (rRNA) sequences reflect the extent of genetic similarity between any two organisms. Closely related taxa have nearly identical rRNA sequences, while microbes that diverged from each other hundreds of millions of years ago exhibit greater levels of nucleotide variation. For the first time it became possible to infer objective dichotomous evolutionary branching patterns that describe taxonomic relationships for cultured microbes. Because the taxonomy reflects the historical evolution of the studied organisms, we refer to these classifications as phylogenies.

At first, the going was very slow. As a graduate student in Woese's laboratory during the late 1960s, I enjoyed tinkering with new technologies. I spent 2 years using "state-of-the-art" two-dimensional oligonucleotide fingerprinting technology to sequence at "warp speed" a small rRNA (5S rRNA) that contained only 120 nucleotides. However, Carl Woese had set his sights on even longer molecules that contained greater levels of evolutionary information within the context of nearly 1,500 nucleotide positions. I decided to take a 30-year hiatus and turned my attention to other, more tractable problems consistent with receiving a graduate degree within my lifetime. Fortunately, Woese persisted. Using cumbersome biochemical techniques to gather partial sequence information from the longer rRNA molecules and primitive desktop calculating machines, the new art of molecular phylogenetics identified nearly a dozen cohesive bacterial phyla. Even before the age of automated DNA sequencing machines, Woese and G. E. Fox made the startling discovery of a third kingdom of life, the *Archaea* (their early branching in the microbial tree of life implied antiquity). The new challenge was to develop strategies for sequencing full-length rRNAs or their genes in order to achieve greater resolution in the molecular phylogenies.

Ten years later, the combination of rapid gene isolation technology afforded by polymerase chain reaction (PCR) and automated DNA sequencing capabilities soon opened another window on microbial diversity. Instead of limiting molecular phylogenetic studies to cultured microbes, Norm Pace used culture-independent techniques to seek novel microbial diversity in naturally occurring, complex microbial communities. Initially relying upon gene cloning and ultimately the use of PCR techniques, Pace isolated and sequenced rRNA genes recovered from environmental DNA samples without the need to culture the microbes. Each of the rRNA gene sequences served as a proxy for the occurrence of a microbial taxon from a sampled environment. By using phylogenetic techniques, microbial ecologists could identify the closest relatives of these environmental rRNA sequences in evolutionary trees that also included rRNA gene sequences from well-characterized, cultivated microbes. When applied to a variety of field samples, a new picture of microbial diversity in natural settings emerged. The number of microbial phyla increased from Woese's description of a dozen bacterial lineages to more than 100 major phyla—most of which do not include cultured representatives. This window on the microbial world produced spectacular discoveries of previously unknown microorganisms, many of which have major impacts on human health and the environment.

Despite the revelations of increased diversity ushered in by molecular technology, these molecular surveys have captured only a fraction of microbial diversity, and they rarely provide estimates of relative abundance for different members in a microbial community. Collector curves that describe the number of different sequences in molecular surveys of complex microbial communities never plateau and rarely describe more than a thousand different kinds of organisms from any one sample. The recovery of certain sequences at very high frequencies from the clone libraries, while others appear as single "phylotypes" scattered throughout the phylogentic tree, suggests that molecular microbiologists have consistently undersampled molecular pools and thus have failed to detect low-abundance taxa. The expense of sequencing full-length rRNA genes (typically four to eight sequence reads per gene, each costing ~$1) has constrained most molecular inventories of microbial populations. Moderate-size surveys that collect 1,000 sequences will detect a small fraction of the 10^8 microbes/liter commonly found in aquatic environments (fewer than one per million cells). These studies detect high-abundance organisms that likely represent rapidly growing populations. They fail to describe the full complement of low-abundance taxa that might play important roles in the response of microbial communities to ecological change. The promise of discovering new phylotypes among the lower-abundance taxa fostered experimental designs where low-resolution procedures—e.g., restriction fragment length polymorphisms—identify putatively distinct clones for DNA sequencing. Unfortunately, this binning procedure comes at the expense of minimizing information about the relative abundance of distinct phylotypes. The occurrence and distribution of the low-abundance taxa remain undersampled and uncharted.

In 2005, the introduction of faster and more efficient "next generation" DNA sequencing technology rekindled my interest in microbial diversity. The ability to sequence hundreds of thousands of DNA molecules for only a few thousand dollars enables molecular sampling efforts that eclipse earlier surveys of microbial population structures by at least 100-fold. Our objective was to comprehensively describe all phylotypes that occurred in various marine environments. We at once succeeded and failed in this endeavor! After sequencing portions of nearly 250,000 PCR-amplified archaeal rRNA genes from diffuse fluids of the crustal aquifer that lies below the underwater volcano Axial Seamount on the Juan de Fuca Ridge, the collector curves reached a plateau. In contrast, the sampling of 750,000 bacterial sequences from the same site failed to comprehensively describe the diversity of the bacterial communities. Using statistical techniques for differentiating between different kinds of bacteria, we estimated that a single liter of seawater from the diffuse flows contained more than 35,000 kinds of bacteria, a completely unexpected level of biodiversity! A relatively small number of sequences represented dominant bacterial populations. A broad distribution of distinct bacterial taxa that represent extraordinary diversity underlies the major populations. These highly divergent, low-abundance organisms constitute a "rare biosphere" that makes up the long tail of rank distribution curves for complex microbial communities. The sheer size of the long tail is enormous, and the genotypic variation of rare microbes eclipses all prior estimates of bacterial diversity. Subsequent studies showed that microbial diversity within soils was considerably greater than that within seawater.

Initial reactions to these results ranged from astonishment to disbelief. If the survey of one million molecules had not fully described the diversity of the microbial community, what level of sampling will be required to fully define microbial diversity within the long tail? And why are there so many different kinds of microorganisms in a liter of seawater? There are no simple answers, but given the enormous number of microbes on the planet and their unrivaled effective population sizes, theory predicts that diversity in the microbial world might be too great to fully measure. Indeed, reanalysis of this initial glimpse into the rare biosphere prompted some statisticians to argue that there are more than 300,000 different kinds of microbes in a liter of diffuse fluids. As for why, we can speculate, but more importantly, the rare biosphere paradigm raises an endless number of questions and hypotheses that might explain the reasons for and the mechanisms responsible for both rarity and diversity in the microbial world. Let the experimentation begin!

FURTHER READING

Bergey DH, Breed RS. 1957. *Bergey's Manual of Determinative Bacteriology,* 7th ed. Williams & Wilkins Co., Baltimore, MD.

Huber JA, et al. 2007. Microbial population structures in the deep marine biosphere. *Science* **318:**97–100.

Pace NR, et al. 1985. Analyzing natural microbial populations by rRNA sequences. *ASM News* **51:**4–12.

Pedros-Alio C, Calderon JI, Gasol JM. 2000. Comparative analysis shows that bacterivory, not viral lysis, controls the abundance of heterotrophic prokaryotic plankton. *FEMS Microbiol Ecol* **32:**157–165.

Quince C, Curtis TP, Sloan WT. 2008. The rational exploration of microbial diversity. *ISME J* **2:**997–1006.

Schloss PD, Handelsman J. 2005. Introducing DOTUR, a computer program for defining operational taxonomic units and estimating species richness. *Appl Environ Microbiol* **71:**1501–1506.

Sogin ML, et al. 2006. Microbial diversity in the deep sea and the underexplored "rare biosphere." *Proc Natl Acad Sci USA* **103:**12115–12120.

Woese CR, Fox GE. 1977. Phylogenetic structure of the prokaryotic domain: the primary kingdoms. *Proc Natl Acad Sci USA* **74:**5088–5090.

Microbes and Evolution: The World That Darwin Never Saw
Edited by R. Kolter and S. Maloy

doi:10.1128/9781555818470.ch5

5

The View from Below

Margaret Riley and Robert Dorit

It's not easy being small. Having to spend a lifetime out of sight and out of mind, blending into the soil and the water, hitching around on large organisms. But this unassuming world of the small is, in effect, what makes life on our planet possible. The complex networks of life are driven by the action of microbes, simultaneously synthesizing and consuming, moving metabolites and energy with relentless, tireless efficiency. At the heart of every critical process in the biosphere—from producing the atmosphere to decomposing wastes—lie microbes. Even if, with our usual narcissism, we care only of what makes human life possible, here too, it is bacteria that enable us to digest food, provide us with vitamins, train our immune systems, and protect us from infection.

And yet, as simple as microbes seem—to some they are mere bags of DNA and proteins—they are full of surprises and contradictions. On the one hand, bacteria, like all other organisms, are engaged in a constant struggle for existence. This struggle is waged

Margaret (Peg) Riley did her academic training at University of Massachusetts-Amherst and Harvard. She was on the faculty at Yale University for many years and moved "home" to UMass Amherst in 2004. Her favorite activity is riding her quarterhorse, Cody, through a snow-filled New England forest; however, she makes time to work with her friend Doc, an African gray parrot, as he expands his science vocabulary. *Robert Dorit* is a professor in the Department of Biological Sciences at Smith College. His education lasted 23 grades and included time at Stanford and Harvard, where he received his Ph.D. in Organismic and Evolutionary Biology and did postdoctoral work in the Department of Cellular and Molecular Biology.

against an environment that can be as inhospitable—too cold, too dry, too exposed to radiation—as it is miserly with certain critical resources—iron, nutrients, vitamins. But the fight is also waged against other bacteria, which must share or compete for the same limited resources. The microbial world, in this sense, is a zero-sum game: one individual's gain is another's loss.

On the other hand, no bacterium is an island. The communities of microbes are deeply interdependent, and every bacterial species depends on one or more other species for survival. The microbial world is organized much like a medieval town: guilds of bacteria cooperate to exploit particular environments and barter and trade with other guilds for the critical resources they require. The consequences of this interdependence are profound, making cooperation and collaboration at least as important as competition in the shaping of bacterial communities.

Had Darwin visited Lilliput during his voyage on the HMS *Beagle,* he would have noted the similarity between macroscopic and microbial life. A nuanced conversation is constantly under way in communities of organisms, be the speakers large or small. The content of that conversation varies from ecosystem to ecosystem, but it always includes a mixture of exhortations, recruiting pitches, and threats. Other essays in this collection deal with the benign and beneficial conversations; here, we focus specifically on the threats and resulting wars waged for dominance. Darwin reveled in the diversity of beak shape, carapace color, and mating behaviors, all the outcome of competition for access to reproduction, food, and space. The microbial world, too, is red in tooth and claw. Microbes wage their miniaturized wars with a diversity of biological weapons and behaviors that help them in their struggle to survive. The weapons range from siderophores, molecules sent out into the environment to scavenge every priceless iron ion, to the bacteriocins, complex and lethal proteins that wreak havoc on close competitors.

At first glance, the meaning of these compounds is anything but subtle: they make life easier for the producers and harder for everyone else. Bacteriocins, for instance, have evolved into finely honed multitasking killing machines. While one domain of the bacteriocin protein recognizes the target cell, another presents a

Trojan horse, which is readily transported into the target cell's interior. Once inside, out springs a third domain and a bloody battle ensues, causing lethal damage to the target cell. Now, in contrast to the H-bomb-like impact of classical antibiotics, which tend to decimate all bacteria in their path, bacteriocins are more akin to guided missiles: targeting one specific strain and sparing the next.

The ability to synthesize a toxic protein raises some vexing challenges not only for the target cells but for the producers themselves. How, after all, can a producer strain keep from becoming the victim of its own lethal weapon? The answer is both surprising and elegant. The producing cell also synthesizes an immunity protein that binds tightly to the killing domain of the bacteriocin, rendering it temporarily inactive. This safety catch will remain on until the bacteriocin is safely inside its intended target cell. At that point, the bacteriocin will arm and set about its deadly business. Not surprisingly, the evolution of bacteriocins and that of their immunity proteins are inextricably linked.

From the producer's perspective, bacteriocins are expensive to synthesize and dangerous to handle. They have evolved to perform a very specific role, clearing ecological space for the producer strain by eliminating close competitors, whose needs and wants most closely resemble their own. Unlike the antibiotics we are used to seeing in clinical settings—coarse, unrefined, indiscriminate, broad-spectrum killers—bacteriocins are precise stilettos. Precisely because they are expensive to synthesize, they have evolved to be of maximum utility to the producers, eliminating only close competitors and ignoring the rest. But an even subtler reality may be driving the evolution of bacteriocin warfare: virtually every bacterial species ever studied depends for its survival on other members of its guild. As a result, laying waste to every living thing in its surroundings strips the producer strain of the ecological partners on which it depends, likely dooming it as well. In contrast, bacteriocins are weapons of a more elegant sort, removing competition but leaving the vital network largely intact.

Not surprisingly, the targets of this elegant warfare also evolve in response. Resistance to bacteriocins is rampant in microbial ecosystems, as the target cells alter their surface receptors, modify their cross-membrane pathways, and evolve machinery to deactivate the bacteriocins. This delicate balance between producer cells and target cells plays out in a surprising minuet that has been likened to the childhood game of rock-paper-scissors. Imagine, for a moment, that the sensitive strain is the rock in our game. A bacteriocin-producing strain emerges and acts as "paper" in our game, defeating the sensitive strain as "paper" covers "rock." Eventually, no sensitive cells survive, and only bacteriocin producers rule. But the emergence of a resistant strain from within the producer population plays the role of "scissors." Little by little, the population shifts from bacteriocin producers to resistant cells: scissors beats paper. Yet in the absence of producers, the cost of resistance is high and the benefits nonexistent. A strain that is no longer resistant—in other words, a sensitive strain—will appear ("rock") and eventually come to dominate, and the cycle begins anew. Around and around, the cycle continues, for millions upon millions of years. No one wins, and no one loses—the cycle simply extends for millennia.

Not every interaction in the microbial world is necessarily subtle. As a species, we have benefitted from the fact that certain bacteria do synthesize small molecules with broad effects. Many antibiotics currently in clinical use are derived from, or inspired by, naturally occurring broad-spectrum antibiotics. At first glance, the existence of these broad-spectrum compounds would seem to contradict the ecological logic we have already discussed. If bacteria are indeed always embedded in an ecological network where one bacterium's waste is another's treasure, why would bacteria evolve the ability to synthesize a compound that kills competitors and potential collaborators alike?

This riddle has deepened of late, as we have developed the capacity to explore microbial ecosystems in situ. We realized early on that the concentrations of these broad-spectrum antibiotics present in the intact microbial systems were way too low to kill surrounding bacteria. At first glance, this made little sense: why evolve a pathway to produce a lethal small molecule and then get all shy about using it? A possible resolution to this paradox has begun to take shape. Antibiotics, it appears, perform a dual role. At high concentrations—the concentrations we use in the clinic—these antibiotics do indeed kill. But at the levels we observe in natural microbial ecosystems, antibiotics act instead as subtle modulators of transcription for specific subsets of genes. Their role may be to whisper, not to yell; to manipulate, not just to kill. At very short distances, producing bacteria may indeed be capable of vacating ecological space in a flash. On the other hand, at the lower dosages most often found in bacterial ecosystems, even broad-spectrum antibiotics may be acting to carve out space for producers by preserving the larger guild rather than by scorching the earth around them. This dual role reflects that same paradox we spoke of earlier: the requirement to cooperate and compete in the very same breath.

Whether viewed from the bow of the *Beagle* or through the lens of a microscope, all life is bound by the same evolutionary rules. All life forms must engage in cooperation, communication, and competition to survive. As environments change, communities mature, and populations fluctuate, the tempo and mode of their evolution may shift, but the rules of the game remain fixed. It is this appreciation for the constancy of the rules that shape the living world that lies at the heart of Darwin's genius.

Microbes and Evolution: The World That Darwin Never Saw
Edited by R. Kolter and S. Maloy

doi:10.1128/9781555818470.ch6

6

Running Wild with Antibiotics

Roberto Kolter

The clock on my nightstand reads 4:37. Once again I am awakened by these dreadful companions, the all-too-familiar ghosts that so often rattle my sleep. Will the grant renewal I submitted recently be funded? How will I pay the salaries of the people training in my lab if I lose that funding? Will I be able to respond to the comments of the referees who liked my latest manuscript but requested an insanely long list of additional experiments? Endless circling thoughts bring mind to consciousness and yes, there is still that long-overdue essay on microbial evolution for the book that I am editing. Such are the circumstances that get me out of bed this morning, like so many other mornings. The dimly lit kitchen is quickly filled with the glow of the laptop's screen coming to life and the aroma of the freshly brewed espresso. After the usual dose of daily news and mindless email catch-up duties, the fingers finally begin to dance on the keyboard and another early-morning writing effort has begun. Why am I willing to lose sleep over some of these painful facets intrinsic to the pursuit of science in the 21st century? Because in the end, there is nothing I would rather be doing than to have the opportunity to explore the unknown. Like everyone else at this stage in my profession—head of laboratory at

Roberto Kolter did his academic training at Carnegie-Mellon University, UC San Diego, and Stanford. Since 1983 he has been a faculty member of Harvard Medical School. A fanatic of food and wine, he enjoys burning those calories off in early morning runs along the Charles River in Boston.

a research university—I have to do battle with the very real monsters of funding and publishing to keep things running. But the rewards of working with bright inquisitive minds in the pursuit of new knowledge keep me going, dissecting the unseen yet fascinating world of microbes. Another espresso, and it's time to lace up my beat-up running shoes.

The clock on the kitchen wall says it is 5:53 a.m. when I open the door to start this morning's run. The brisk air, the gray sky, and the scheduled slow easy pace for today's 6-miler are propitious for the mind to wander. By the time I have warmed up and reach the banks of the Charles River the mind is wandering indeed. The peaceful water surface of the windless morning contrasts sharply with the storm inside as I recall the morning headline. Another young and healthy high school athlete has succumbed to a fulminant MRSA infection. MRSA, methicillin-resistant *Staphylococcus aureus*, epitomizes the present-day problems confronting infectious disease physicians. Bacterial infections, for eons the scourge of human populations, seemed conquered by the middle of the 20th century. The antibiotic era, stemming from the first clinical use of penicillin in the 1940s, along with vaccination and sanitation, provided a sense of confidence in medicine that infectious diseases had been conquered. As U.S. Surgeon General William Stewart declared in 1967, it was time to "close the book on infectious disease." But sadly, today infectious diseases continue to be the predominant cause of death worldwide. And the increase in the incidence of incurable infections, exemplified by those caused by antibiotic-resistant bacteria such as MRSA, makes the future seem bleak. The antibiotic era may indeed be agonizingly short. Bacterial infections that were easily treatable 20 years ago are nowadays becoming increasingly difficult to control. As I pound the pavement, I am captured by the pain of the parents' despair at the news of the death of their healthy child at the hands of an invisible bacterium. With all that scientific knowledge and all those antibiotics that have been developed, how come their child could not be saved? This disturbing thought bounces rhythmically in mind, burdening every footfall of my run. Does not make sense.... Does not make sense....

As I cross the 3-mile mark, 6:29 a.m. on the wristwatch, the mind remembers that the problem of antibiotic-resistant bacteria does, sadly, make perfect sense. It's evolution at work. As a biological phenomenon, antibiotic resistance makes sense, proving the wisdom of Theodosius Dobzhansky, who in 1973 remarked, "Nothing in biology makes sense except in the light of evolution." To make sense of antibiotic resistance we must see it for what it is, evolution in action. Right in front of our very own eyes, evolution is happening in real time in time scales much shorter than our own lifetimes. But...is not evolution supposed to be so slow that it is difficult to observe it while it happens? Perhaps so with plants and animals. But not so with bacteria. Bacteria are able to grow quickly to remarkably large population sizes and have an uncanny ability to exchange genes across different species. These features render them capable of extremely rapid evolution. Consider, if you will, your body. All in all, there are at least 10 times more bacterial cells living on our skin, mouth, and gut than there are human cells in our bodies. That is hundreds of trillions of bacterial cells constantly evolving in every single human. The evolution of antibiotic resistance is an example of what is often referred to as the "Red Queen effect." The name is adopted from the classic Lewis Carroll work *Through the Looking Glass*. After Alice runs through the Garden of Live Flowers, the Red Queen tells her, "It takes all the running you can do, to keep in the same place," conjuring up the image of running on a treadmill. I, for one, am much happier with the feeling of running along the Charles River than the captured feeling I get from the vicious cycle of running on a treadmill. Speed up the treadmill and you had better speed up your running—just to keep in the same place—or face falling off the machine. When applied to evolution, the Red Queen concept is simple: evolve fast enough to meet changing conditions or become extinct. As we have rapidly introduced antibiotic use during the last 60 years, bacteria countered by rapid evolution of resistance; they certainly have not become extinct.

A glance at the watch tells me it's 6:47 a.m. when I reach my driveway. I've picked up the pace in the second half, perhaps to avoid thoughts of extinction. In running, the pace defines the race. Stretching my tightened hamstring muscles, I dream up the allegory of running to explain the rapid evolution in antibiotic resistance we are witnessing. In competitive running, there is the wide range of distances, from the 100-meter dash to the marathon—over 42,000 meters—and beyond. Runners must take dramatically different approaches to racing each distance. Sprinting for 100 m requires an explosive effort from the start and sustained maximum effort that—for the best of runners—is over in less than 10 seconds. Marathon running is another world. Start too fast and you will quickly burn out. Better learn to warm up slowly, gently increase the pace, and finish strong.

It seems that as a species, we humans approached antibiotic use as if it were a 100-meter—sorry, 100-year—race against bacteria. We came off the blocks with a bang, as evidenced by the remarkable cures observed during the early days of penicillin. The first 20 years of the race saw explosive acceleration in the field. Hundreds of new antibiotics were discovered and put into use. We looked unbeatable and we got cocky. We began to use antibiotics widely and indiscriminately in the clinic, and used them also to promote the growth of livestock. It is estimated that today worldwide antibiotic production surpasses 100,000 tons annually, with well over half of that being used not to treat infections but as supplements in livestock feed. It is no surprise. Antibiotics are big business; some $25 billion is made yearly from their sales worldwide. It would have seemed that we could cruise the rest of the race for an easy and permanent victory against pathogenic bacteria. But early on in the race, bacteria began to show signs of gathering momentum in their favor. Yes, they might have faltered a bit at the start, when they looked as if they had stumbled coming out of the blocks. But soon enough, antibiotic-resistant pathogens began to appear in the clinical setting. Bacteria had population sizes on their side, and evolution began to pick up the pace. By the 1960s many multiply resistant pathogens began appearing in clinics all over the world. Today, nearing the 70-year mark of the antibiotic era, we are running neck and neck with the bacteria. We still can control many infections with antibiotics, but there are ominous reports of some pathogens now being resistant to all known antibiotics. To make matters worse, we are losing steam: the rate of new antibiotic discovery has diminished dramatically in the last decade. The way things are going, it looks like the bacteria will win this 100-year race.

There could have been a different strategy in approaching antibiotic use. We could have thought of it not as a 100-year race but more as a marathon, a 42,000-year race. We could have done all the training necessary and gone ahead and discovered and developed all the antibiotics that we have today but simply shown restraint in the pace at which we used them, restricting them to clinical use and only when absolutely necessary. Then, no doubt, we would have the prospect of antibiotic resistance becoming a problem but along thousands of years rather than in less than a century. Things look grim now, but there is still time to bring down the pace of this race. And there are signs that we are beginning to do so. Some countries are enacting much stricter controls on antibiotic use, and scientists are hoping to develop novel antibiotics that, because of their very nature, do not lead to the rapid evolution of resistance.

Enough stretching for now: it's almost 7:30 and I need to shower and eat breakfast before I get on my bike and head on to the lab. We have to keep learning more and more about what is the natural role of these compounds that we call antibiotics. You see, the majority of the antibiotics in use today are compounds that bacteria produce naturally. Surprisingly, while we know that they are indeed capable of killing bacteria in the laboratory and in patients during infections, we do not really know what their function is in their natural setting. In some cases, we have pretty convincing indications that bacteria do use these compounds to fend off other bacteria. But we have also found equally convincing indications that at sublethal concentrations these compounds can be used by bacteria to send important messages to other bacteria. In some cases these messages cause the receiving bacteria to disarm themselves of their virulence. But due to the fact that the receiving bacteria are not being killed, there is no selective advantage in the survival of resistant mutants. In such settings the evolutionary process should not favor the spread of resistance. Thus, as we gain a greater understanding of the ecological role of antibiotics, we should be able to devise strategies that will allow us to slow down the pace of our race with the bacteria, with the concomitant result that the race can last far longer than the 100-year sprint we are so close to losing. The process of gaining that understanding is filled with surprises—never a dull moment. New ideas constantly arising, challenges that need to be met, young minds that need to be trained without forcing them closed. The race is long and slow, just the way I like it. I'll keep at it for as long as I can, despite all the pains of grant proposals and referees' comments. And in the meantime, I'll keep trying for that elusive Boston qualifier...some day. If not this year, maybe next.

Note added in proof: The author recently achieved that elusive Boston qualifying time at the Bay State Marathon.

Microbes and Evolution: The World That Darwin Never Saw
Edited by R. Kolter and S. Maloy

doi:10.1128/9781555818470.ch7

7

Antibiotic Resistance

Diarmaid Hughes

Jim Henson, the creator of the popular children's series called "The Muppets," started feeling tired and had a sore throat. He felt like he was fighting a cold over the weekend and thought he'd get better soon. However, by Monday he began to have trouble breathing and started coughing up blood. He was quickly admitted to the hospital and given multiple antibiotics. However, less than 24 hours later his lungs, kidneys, heart, and blood-clotting systems failed, and he died. Less than 24 hours after he felt sick, Jim Henson was dead from an infection with *Streptococcus pneumoniae*!

This story is repeated every day around the world. Sometimes it is *Streptococcus*, sometimes it's methicillin-resistant *Staphylococcus aureus* (MRSA); other times it's another one of the many different types of bacterial infections. Antibiotics are supposed to protect us from infections with these pathogenic bacteria...why didn't they save Jim Henson, and why is this story becoming so common?

What Are Antibiotics and How Do They Work?

Antibiotics are small molecules that inhibit important functions in bacterial cells. For example, some interfere with the building of

Diarmaid Hughes was educated in Genetics at Trinity College Dublin, then in 1985 moved to Uppsala, where he deepened his interest in microbial growth physiology and evolution. He enjoys working to the music of Mozart and Bach and finds relaxation on the west coast of Ireland facing the Atlantic winds.

new cell walls, causing cells to lyse when they attempt to grow; others interfere with enzymes that wind and unwind DNA, causing chromosomes to break; finally, others interfere with the ribosome, preventing bacteria from making new proteins. Typically, each class of antibiotic has a single major target in the bacterial cell. This target is usually an enzyme to which the antibiotic binds, thus preventing it from carrying out its important function. Most antibiotics act as "a spanner in the works," preventing enzymes from going through their normal cycle of activities and conformational changes.

The development of antibiotics was one of the greatest medical advances of all time. Half a century of research and development by pharmaceutical companies has given physicians about two dozen classes of antibiotics to deal with most bacterial infections that trouble us. Antibiotics eliminated the fear of premature death from tuberculosis and the pain of urinary tract or middle ear infections. Most importantly, antibiotics made it possible to perform complicated surgery without a high risk of death due to infection. Performing heart transplants, or treating the victims of serious automobile accidents or battlefield casualties, would today be unthinkable without antibiotics to reduce the threat of serious bacterial infections.

We now take antibiotics for granted. So, what's the problem, and what has it got to do with Darwin, species, and evolution? Antibiotics are losing their effectiveness at an alarming rate because of the evolution of antibiotic-resistant pathogenic bacteria. Although there is a very large social aspect to this problem (our frequent inappropriate use of antibiotics), I focus in the remainder of this essay on the genetic and evolutionary aspects of antibiotic resistance, and how an unintended global experiment in selection has taught us about genetic variation and evolution in the bacteria.

Some Bacteria Are Naturally Resistant to Particular Antibiotics

Antibiotics are not universally effective against all bacteria: each antibiotic has a biological spectrum of activity, and there are three main factors that limit this spectrum. An antibiotic may fail to reach its intended target, for example, because it cannot cross the cell wall and get inside the bacterium. This is a major reason why many antibiotics are ineffective against gram-negative bacteria, and why pharmaceutical companies have devoted great research efforts to developing variants of successful antibiotics with different activity spectra. A second cause of intrinsic resistance can be an absence of the target. It's unusual for a pathogen to completely lack the type of important function usually targeted by antibiotics, but in some cases the enzyme carrying out that function differs significantly in structure in certain bacteria, and thus the antibiotic does not affect the alternative enzyme. The third factor is that some bacterial species carry genes coding for resistance mechanisms possibly because those species evolved in environments where they were exposed to naturally occurring antibiotics. Such resistant bacteria are a source of resistance genes that can be transferred to other bacteria by a process called horizontal gene transfer (HGT). When we began using antibiotics to eliminate troublesome pathogens, we unfortunately failed to appreciate the extent of genetic diversity in the bacterial world and the significance that HGT would have for the spread of resistance to pathogens.

How Susceptible Bacteria Become Resistant to Antibiotics

Genetic alterations can cause previously susceptible bacteria to become antibiotic resistant. There are many different genetic pathways by which bacteria can evolve antibiotic resistance, but they can be broadly divided into two classes: (i) those where genetic alterations reduce the effective interaction between the antibiotic and its target in the bacterial cell and (ii) those where genetic alterations reduce the chemical effectiveness of the antibiotic molecule itself. A second important aspect to consider in the evolution of antibiotic resistance is the source of the genetic alteration responsible for the resistance. In some cases the source is intrinsic to the targeted pathogen—for example, a mutation may occur in the bacterium that reduces its susceptibility to the antibiotic. This is a classic example of Darwinian evolution: in any large bacterial population we can expect that there will be mutants, some of which will have resistance to the antibiotic, and that individuals with these properties will increase in frequency under antibiotic therapy. In other cases the resistance may be conferred on the target pathogen by its acquisition of foreign DNA by HGT. There are several different mechanisms of HGT, and some of these are also introduced in the following section. The global significance of HGT in providing the raw material for bacterial evolution in general, and antibiotic resistance specifically, is one of the major lessons we are still learning from our use of antibiotics. A third issue is the relationship between acquired antibiotic resistance and fitness of the pathogen. Because most antibiotics target essential and highly conserved functions, most alterations that make bacteria resistant also reduce their fitness in important environments—for example, by causing slower growth or reduced virulence. When the antibiotic is present this is not a problem for the resistant bacteria: they will have a much higher fitness level than susceptible bacteria. However, in environments without antibiotic selection the resistant mutants may suffer reduced fitness relative to that of their susceptible relatives. This could result in the loss of resistant mutants from the population, but in practice it often results in the selection of fitness-compensatory mutations in the resistant bacterial population. Thus, after the initial acquisition of resistance, the resulting reduction in fitness frequently drives the further evolution of the resistant bacteria, leading to the evolution of a new genetic entity: a bacterium resistant to the antibiotic and also genetically adapted for growth in the presence or absence of the drug.

A final issue to consider is why antibiotic resistance has become such a major medical problem. After all, the genetic alterations that make bacteria resistant to antibiotics are rare and occur in individual bacteria. The factor that makes these resistant bacteria such a major problem is that individual bacteria with these phenotypes—antibiotic resistance combined with high fitness—have been effectively selected by our use of antibiotics and are replacing populations of susceptible bacteria. We humans have for decades been carrying out a largely unsupervised global experiment in bacterial evolution, and we are now suffering the consequences.

Target-Focused Resistance Mechanisms

Mutation of the target

One way to evolve antibiotic resistance is by mutating the gene coding for the target of the antibiotic. There are many examples of single mutations causing resistance. For example, rifampin is an important first-line drug for the treatment of tuberculosis caused by infection with *Mycobacterium tuberculosis*. Spontaneous mutations in the gene coding for one of the protein subunits of RNA polymerase, that change a single nucleotide that produces a single amino acid substitution in the protein, are sufficient to cause high-level resistance to this antibiotic. In cases where the target gene is mutated, resistance usually occurs because the mutant target protein has a reduced affinity for the antibiotic.

Drastic alteration of the target

Sometimes single mutations are not sufficient to make a target resistant. Because specific individual mutations are rare (typically $\leq 10^{-9}$ per generation), one might think that drastic alterations involving multiple changes would be so rare as to be irrelevant to resistance evolution. Unfortunately, this is not always the case. For example, penicillin-type antibiotics provide safe effective treatment for infections by *Streptococcus pneumoniae* (the pathogen that killed Jim Henson). Penicillins target cell wall building blocks called penicillin binding proteins (PBPs). In *S. pneumoniae* resistance is caused primarily by the acquisition of PBP genetic material from other bacterial species. These foreign genes code for proteins that carry out the normal functions of the original PBP but are so divergent in sequence that the PBPs they encode are resistant to the antibiotic.

Protection of the target

Another mechanism of target-focused resistance is by the production of a molecule (usually a protein) that binds to the target without inhibiting its normal function and protects the target from the antibiotic. If the bacterium has the ability to produce a target-protecting protein, then it will display intrinsic resistance; otherwise, it may acquire this mechanism by HGT. For example, target-protecting proteins that bind to DNA topoisomerases, the target for quinolone antibiotics like ciprofloxacin, often abbreviated to "Cipro," and that prevent drug binding to the target are one of the causes of increased resistance to this very important group of antibiotics.

Replacement of the target

The major cause of resistance to vancomycin, the antibiotic of last resort for several important infections, including MRSA infections, is by the acquisition by HGT of a set of genes that remove the normal cell wall target of the antibiotic and replace it with an alternative cell wall building block which carries out the same function but is not sensitive to vancomycin. Target-focused mechanisms generally work by reducing the sensitivity of the target to inhibition by the antibiotic.

Drug-Focused Resistance Mechanisms

Drug inactivation

A resistance phenotype can also be generated by any mechanism that inactivates the antibiotic. There are many examples where a gene encodes an enzyme that modifies an antibiotic such that it loses its antibacterial activity. The best-known example is the inactivation of penicillin-like antibiotics (β-lactams) by enzymes called β-lactamases that cleave the antibiotic. These β-lactamases are typically encoded by genes on transmissible plasmids that can be rapidly transferred between pathogens. Attempts by pharmaceutical companies to combat β-lactamases through the development of novel variant penicillins resistant to the enzymes has led to an evolutionary arms race: β-lactamases have coevolved with the new antibiotic variants, and we now have a bewildering variety of penicillin-like drugs and pathogens expressing β-lactamase-type enzymes that cleave them.

Drug efflux

Bacteria have evolved mechanisms to import nutrients from their surroundings and efflux toxic substances out of the cell. In some species intrinsic resistance to an antibiotic is due to the presence of efflux systems that efficiently pump the antibiotic out of the bacterium. Efflux causes resistance by reducing the internal concentration of the antibiotic below the level required to inhibit target function. In bacteria that are susceptible to an antibiotic, a weak intrinsic efflux activity can often be increased by mutations that cause overexpression of the pump or by mutations that change the specificity of pump. In other cases, bacteria have acquired a novel efflux pump activity, and antibiotic resistance, by HGT.

Drug influx

Although not as well documented as efflux pump mutations, there are mutations affecting membrane proteins that reduce the rate of antibiotic entry into the cell. Each of these drug-focused resistance mechanisms works by reducing the concentration of antibiotic that can reach the drug target.

What Has Antibiotic Use Taught Us about Natural Variation, Selection, and Evolution of Bacteria?

Darwin proposed that species contain within them variants that can increase or decrease in frequency subject to selection. We now appreciate that population bottlenecks and genetic drift also play an important part, and we know from molecular genetic studies the nature of the genetic variants. We could have predicted from Darwinian theory that the use of antibiotics might have had two different effects on bacterial species: (i) that some targeted species might be driven to extinction (if they did not contain within them variants resistant to the antibiotic) and (ii) that the phenotypic and genotypic composition of other species (whether targeted intentionally or not) would be altered by our selection of variants present in those species that happened to be more resistant to the antibiotic.

What Darwin did not specifically address, and we did not appreciate at the beginning of the antibiotic era, was the nature of bacterial species and the rate and extent of HGT. Although arguments continue, most biologists agree that bacteria can be usefully grouped into species defined in terms of core genome sequences and phenotypic properties. This definition, although much fuzzier than the species definition for animals and plants, for which productive sexual relations set stricter boundaries, is nevertheless of practical use. The existence of HGT does not mean that species definitions are meaningless or that genetic barriers do not exist. Rather, it means that the barriers are to some degree porous, and we are currently learning the hard way where some of these pores are and how large they might be. HGT can have a huge impact on the tempo and direction of bacterial species evolution. Thus, our extensive use of antibiotics is gradually revealing to us something of the genetic interconnections that exist throughout the bacterial kingdom. We would have predicted that genetic variants existed in targeted pathogen populations and that antibiotic use would select for an increased frequency of resistant variants. What was not appreciated until recently was that a genetic resistance determinant existing *anywhere* in the bacterial kingdom would be able to find its way into the targeted pathogen population.

The mechanisms of HGT include transformation by naked DNA, transduction by bacterial viruses carrying random pieces of genetic information from the last cell they infected, and conjugation of plasmid vectors carrying resistance determinants that can transfer between different bacterial species. Although foreign DNA may initially be poorly expressed in the pathogen, as long as it confers some small selectable advantage on the pathogen, then further adaptive evolution by mutation and/or gene amplification can rapidly optimize its expression. The evidence we now have suggests that most of the bacterial domain shares a single very large gene pool. We still know far too little about this gene pool to be able to make reasonable predictions concerning what might happen next. We need to learn more about the total genetic information content of this pool, the rates of genetic transfer between closely and distantly related species, and whether there are major currents of gene movement. At the moment all we can predict with reasonable certainty is that if we use an antibiotic against a susceptible pathogen population we will, with very high probability and within a short time, select out a variant that either has mutated its way to resistance or has acquired by HGT a resistance determinant that had previously evolved elsewhere.

Where Do We Stand Today?

Regarding the medical usefulness of antibiotics, we are clearly heading towards very big trouble. The problem of resistance is worsening and will eventually be noticed by those in political power when it severely restricts certain medical options. The pharmaceutical companies, however, have largely withdrawn from the antibiotic development field because it is currently not seen as profitable. Even if a dire medical situation eventually forces a resolution of the profit issue and stimulates a major development effort, the time it will take to find a new generation of useful antibiotics is likely to be on the order of one to two decades. If we do eventually develop new antibiotics against which there is no preexisting resistance in targeted pathogen populations, then it will be imperative that we learn from the lessons of the past and design treatment strategies with the twin aims of developing effective therapies and maintaining a long useful life span for the antibiotics.

Regarding our understanding of genetic variation, HGT, and evolution in the bacterial kingdom, the widespread use of antibiotics has revealed a level of genetic adaptability and interconnection that is truly breathtaking. Significant new functions that can be acquired in a single step by HGT raise interesting issues concerning the process of speciation and what it means to define a bacterial species. It is hard to imagine that we scientists would ever have sought, much less been granted, ethical permission to carry out the global experiment in bacterial evolution that has been inadvertently conducted through the widespread use of antibiotics over the past decades. The early hope that antibiotics would eradicate infectious diseases has turned out to have been hopelessly naïve and wrong. Crucially, we failed to realize that the use of antibiotics as a weapon against a few specific troublesome pathogens would actually engage us in a long-term war against the entire bacterial world. This is rather unfortunate because not only is the bacterial world overwhelmingly strong in terms of total number of species and total biomass, but also it contains what is probably the greatest diversity of chemical and enzymatic activities on the planet. In retrospect, we have learned a lot about the biological world and about genetic variation and evolution, but we should probably have targeted our enemy more carefully.

Microbes and Evolution: The World That Darwin Never Saw
Edited by R. Kolter and S. Maloy
©2012 ASM Press, Washington, DC
doi:10.1128/9781555818470.ch8

8

Bacteria Battling for Survival

Thomas M. Schmidt

I majored in biology as an undergraduate student, but for me, the life sciences provided a collection of fascinating but mostly disjunct set of facts and observations. Oh sure, I saw the beauty in Darwin's theory of evolution and delighted in the explanatory power of the central dogma of molecular biology that explained the flow of information between DNA, RNA, and protein, but I studied the basics of plant and animal biology separately, without building an intellectual framework that provided meaningful links between those large bodies of knowledge.

My view of biology changed abruptly when I entered graduate school. To the dismay of my graduate advisor, my undergraduate curriculum did not include a single course in biochemistry or microbiology. I hadn't avoided these subjects; it was just that to me, a kid from a Midwestern steel town and the first generation to attend college, biology was primarily the world of plants and animals. Wow, was I wrong! My first courses in microbiology initiated a fundamental revision of my view of the biological sciences: in the world of microbes I first united the fields of biochemistry, genetics, physiology, ecology, and evolution. I was

Tom Schmidt received his Ph.D. from The Ohio State University and conducted postdoctoral research at Scripps Institute of Oceanography and Indiana University. He has been a faculty member at Michigan State University since 1993 and spent several summers in Woods Hole, MA, as director of the Microbial Diversity Course. Outside of the lab, he is known to assume his alter ego as a drummer in a local rock band.

thrilled, and I decided to pursue a doctoral degree studying single-celled organisms. This was new territory for me and for my family—there were not lots of microbiologists to serve as role models in a steel town. However, everyone knew that bacteria cause disease, and so it sounded like a noble pursuit. I wasn't specifically interested in pathogens but was attracted to the possibility of studying the ecology of microbial life. After all, microbes had inhabited Earth for almost 4 billion years and are still the most abundant forms of life on the planet.

The problem with my plan was that while millions of microbes can be seen in a drop of water or a gram of soil, little was known about the majority of these organisms. Conventional microbiological studies are based on the study of microbes that can be grown in the laboratory, and incredibly, less than 1% of the microbes seen in nature are readily cultivated in the laboratory. It's hard to study the ecology of a microbial community when the majority of players are unknown! So I conducted my graduate studies with a few of the cultivated microbes that were distinguished by their exotic lifestyles, like those which harvest energy from the oxidation of hydrogen sulfide (a gas more toxic to us than cyanide) or light-producing bacteria that are symbionts in nematodes. Fortunately, while I was mastering the tools to study microbes that could be grown in the laboratory, biology was shaken by the discovery of a new form of life. Serendipitously, that discovery also set the stage for a new era in microbial ecology.

Carl Woese, a molecular biologist-turned-evolutionist, recognized that for biology to advance, it needed an evolutionary framework that included microbes. After all, it was in the microbial world where evolution shaped the biochemical machinery for information processing and energy generation that underpins all of life. To generate an evolutionary "family tree" that included all forms of life, Woese compared the sequences of bases found in ribosomal RNA (rRNA)—a central component of the ribosome, which is where the sequences of bases in messenger RNA (mRNA) are translated into proteins. As organisms evolve, they accumulate mutations in the rRNA-encoding gene, and Woese reasoned that the more similar the sequences of rRNAs from any two organisms, the more closely they were related to one another. It was elegant and dramatically powerful logic. However, gathering the data to test his model was not trivial, and he labored for more than a decade to generate the information he needed to make these evolutionary comparisons.

As he compared the rRNA sequences, Woese uncovered a previously unrecognized form of life now known as *Archaea* (Fig. 1).

Originally thought to occur only in physically extreme environments like deep-sea hydrothermal vents or the salt-saturating conditions in the Dead Sea, we now know that *Archaea* are ubiquitous in nature, including in the gastrointestinal tracts of approximately half the human population. Woese's "Big Tree" also revealed that the evolutionary diversity was present in the world of microbes, with plants and animals occupying a small corner of the genetic diversity of life. We now had the evolutionary framework that Darwin had envisioned in *On the Origin of Species*, and its impact on biology has been profound.

The new view of evolutionary relationships laid the groundwork for a new era of microbiology, one that included the capacity to study microbial communities without growing them in the laboratory. By determining the sequences of rRNA-encoding genes in DNA extracted directly from nature, we could now determine the composition of microbial communities and study their dynam-

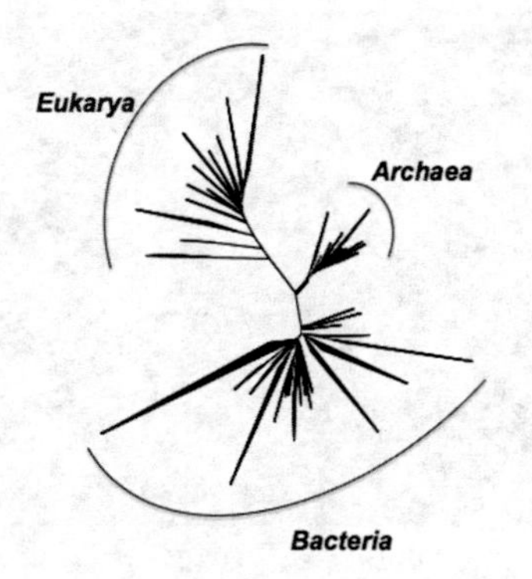

Figure 1 The three domains of life as defined by a comparative analysis of the sequence of rRNA-encoding genes. Plants and animals occupy a small corner of the eukaryotic domain of life. doi:10.1128/9781555818470.ch8f1

ics. Would Darwin's idea of the environment selecting for the fittest of plants and animals also apply to microbes?

During my undergraduate days, "survival of the fittest" conjured up notions of the fastest and fiercest carnivore winning the race for prey, or male birds with the most elaborate display of plumage winning the attention of potential mates and leaving more offspring. This would have to be adjusted for the microbial world, where nutrition comes primarily through diffusion from the environment and reproduction is primarily asexual. So what exactly constitutes the "fittest" in the microbial world?

The goals of life are the same whether we are considering buffalo or bacteria: survive and reproduce. For bacteria, survival means weathering periods when no food is available—which is the majority of the time—and reproducing when conditions permit. Rapid growth is advantageous in environments where nutrients are frequently available, for instance, in the gastrointestinal tracts of animals, but is fast growth always the optimal strategy? We began thinking about the slow-growing bacteria that take days or weeks to form tiny colonies which are barely visible to the naked eye (Fig. 2).

What is their strategy for success? How do they compete against the fast-growing microbes?

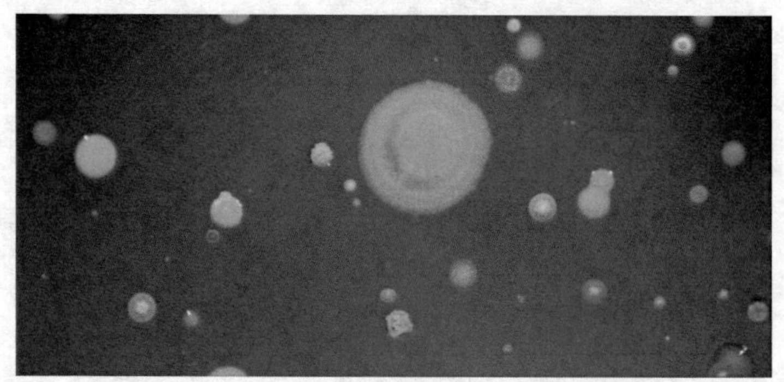

Figure 2 Bacterial colonies growing on the surface of a nutrient medium. Rapidly growing bacteria typically form large colonies within a day or two, while weeks are required before some of the small colonies formed by slow-growing bacteria are visible. doi:10.1128/9781555818470.ch8f2

It is here that we began to think about trade-offs. Darwin had observed trade-offs in the size and power of the beaks in his studies of finches in the Galápagos Islands. Some finches had evolved delicate beaks that allowed them to manipulate and eat small seeds present on some islands, but with this improved capacity for handling small seeds came the "expense" of being unable to break the hulls of larger, more rugged seeds—those were handled by finches that evolved more powerful beaks. Trade-offs between power and efficiency are common not only in the biological systems but in the physical sciences and engineering. Consider the design of internal combustion engines: the horsepower that provides for rapid acceleration inevitably leads to an engine that delivers lower fuel efficiency. With regard to survival of the fittest in the microbial world, perhaps considering a trade-off between power and efficiency would provide some useful insight.

Power can be defined in terms of the capacity of microbes to grow explosively, while efficiency can be captured in the number of progeny a microbe produces from a fixed amount of nutrients. In environments where nutrients are seldom abundant, for instance, deep soils or at depths where light can no longer reach in oceans and lakes, the environment may select for microbes able to use resources efficiently and leave the most progeny rather than those that grow most quickly.

Research from my laboratory group supports a model for trade-offs in the central flow of information in cells—the translation of RNA into protein by ribosomes. This is where cells spend the majority of their energy, linking amino acids together to form a functional protein. And so if there is selection for efficient energy utilization, it makes sense for selection to act on the ribosome. We have found basic differences in the rate at which ribosomes synthesize protein: the rapidly growing bacteria synthesize proteins much faster, but they appear to do so less efficiently. This offers the prospect of a fundamental mechanism that could help explain the distribution of microbes in natural and managed ecosystems. Although it was not immediately obvious how Darwin's ideas would apply to the microbial world, we are beginning to understand some basic differences in microbes and how the environment may select for "survival of the fittest"—even in the world of single-celled organisms.

FURTHER READING

Dethlefsen L, Schmidt TM. 2007. Performance of the translational apparatus varies with the ecological strategies of bacteria. *J Bacteriol* **189:**3237–3245.

Lee ZM-P, Bussema C III, Schmidt TM. 2009. *rrn*DB: documenting the number of rRNA and tRNA genes in *Bacteria* and *Archaea*. *Nucleic Acids Res* **37:**D489–D493.

Stevenson BS, Schmidt TM. 2004. Life history implications of rRNA gene copy number in *Escherichia coli*. *Appl Environ Microbiol* **70:**6670–6677.

Weiner J. 1994. *The Beak of the Finch: A Story of Evolution in Our Time.* Random House Inc, New York, NY.

Woese CR, Kandler O, Wheelis ML. 1990. Towards a natural system of organisms: proposal for the domains *Archaea, Bacteria,* and *Eucarya*. *Proc Natl Acad Sci USA* **87:**4576–4579.

Microbes and Evolution: The World That Darwin Never Saw
Edited by R. Kolter and S. Maloy

doi:10.1128/9781555818470.ch9

Phage
An Important Evolutionary Force Darwin Never Knew

Forest Rohwer

At the end of college, having read most of the classical evolution books and taken a number of courses on the subject, I was pretty convinced that evolutionary studies were effectively over. There really wasn't anything that appeared to be terribly interesting, and most evolutionists were stuck "proving" evolution by sequencing genes. Nothing as a graduate student in molecular immunology really changed my view of evolutionary biology or piqued my excitement. The only related field that I kept up with was the origin-of-life studies by Gerald Joyce and colleagues. Then, in the early 2000s, I started hanging out with microbiologists. I recall hearing Farooq Azam give a talk about 10 million phage per milliliter of seawater. (Viruses that attack bacteria are known as bacteriophage, i.e., eaters of bacteria, or simply, as in this essay, as phage.) Here was something cool!

Forest Rohwer is a professor of biology at San Diego State University, CA. He received his Ph.D. from the SDSU/UCSD Joint Doctoral program in molecular immunology, studying interleukin-2 signal transduction. He then moved to Scripps Institution of Oceanography and developed metagenomic approaches to study marine viruses with Farooq Azam. In 2002 he started his own lab at SDSU, where the main areas of study are metagenomics and coral reef microbiology. He likes spoiling his daughter Willow, drinking San Diego microbrews, fishing, SCUBA diving, and reading in the bathtub.

It was Anca Segall who suggested that we sequence the genome of one of these marine phage. By pinching money from various grants, we were able to piece together the genome of one phage that had been characterized at the Scripps Institute of Oceanography, locally known as Roseophage SIO1. The genome consisted almost entirely of genes that were unrelated to anything that had been reported before. There was one region, however, that was obviously a module involved in replication of the phage DNA, including a DNA polymerase, helicase, and endonuclease. This observation led me to the classical work by David Botstein on modular evolution and the more recent work by Graham Hatful, Roger Hendrix, and Jeff Lawrence on mosaic evolution. Botstein showed that different phage could recombine to form new chimeric phage. He had proposed that "spacer" regions between the genes were hot spots for these exchanges. His model turned out to be wrong, but the effect was the same. Modules of genes that work well together stay together. The genomic sequences showed that there were no linkers. Instead, foreign genes were spliced directly into working modules. Hence, the genome is a mosaic of genes from different phage.

It became clear that phage represented the ultimate mash-up of selfish genes producing hopeful monsters—that is, groups of genes that promote their own propagation but which can often result in dramatic changes for the host that acquires them. Furthermore, these monsters exist in numbers that no one had ever considered before. Suddenly, the impossible was almost a certainty. Imagine 10^{31} phage on our planet, each with a half-life of approximately 10 days. To make replacement phage, they need to infect host cells, which contain other, integrated phage. And of course when this happens, both phage try to get out of the doomed host cell as soon as possible. Each phage tries cutting up the other's genomes, as well as the host genome. Suddenly DNA strands are floating all over the place, getting spliced together to generate new phage chimeras. These chimeras are then released into the environment for natural selection to do its culling. The number of potential exchanges indicates that phage participate in a remarkably rapidly evolving and enormous global pool of genetic information—something I like to call the "virome" because it includes all viruses. I wanted to focus my attention on efforts to measure and characterize this gargantuan gene pool.

Determining the global virome had a number of significant challenges, the first problem being that most of the microbial hosts are not easy to culture. Even if all of the hosts were tamed to grow in the lab, it still would have been essentially impossible to isolate enough phage to really make a dent in our efforts to catalog global viral diversity. The second major challenge to determining the size of the global virome was that phage do not have any signature genes like the 16S ribosomal DNA (rDNA) used to characterize *Bacteria* and *Archaea*. Although at the time this seemed like a major hurdle, it turned out to be a lucky break because it sent us down another avenue for determining the diversity of uncultured viruses. This approach, which eventually became known as viral metagenomics, was to randomly clone pieces of uncultured viruses and then sequence them.

The first metagenomes were viral communities taken from seawater off the coast of San Diego. Their most interesting property was the observation that <20% of the sequences had any significant similarity to previously known sequences. This generality was still true 7 years later, even though the databases had grown exponentially since 2002. In subsequent metagenomes from microbes, nearly 80% of the genes were found to be similar to other sequences catalogued in GenBank. This means that most of the uncharacterized genomic diversity on our planet is in the viruses. And this diversity is enormous! Based on the metagenomic analysis, we estimate that there are ~2.5 billion novel phage-encoded genes yet to be discovered on Earth (and I sincerely hope that there are even more on Europa).

The massive genetic potential contained in the global virome has profound implications for evolution because of horizontal gene transfer. One example that directly influences humanity is the production of new pathogens. Many common microbial pathogens, including *Staphylococcus, Streptococcus, Salmonella, Vibrio cholerae,* and many others, are actually sick themselves. These poor, innocent bacteria are infected with a phage which has turned them into a disease-causing pathogen. Moreover, the phage themselves are sick, because they are infected with a selfish gene encoding a toxin.

The phage-encoded toxins are the real culprits because they cause most of the disease symptoms. In effect, they have forced the bacteria to turn some animal (like ourselves) into a factory to produce the pathogenic bacteria. As we pour our body fluids into the environment, the bacteria escape and then are blown up to release the phage and their associated toxin genes. In 2001, Mya Breitbart, Veronica Casas, and I decided to see how common the toxin genes that cause diphtheria, anthrax, cholera, and whooping cough were in the virome around San Diego. When we tested naturally occurring viromes for the toxin genes, we found that they were very common. Ten percent of the samples had at least one of the toxin genes. Given this environmental pool of toxin genes, we will never be able to eliminate these potential genetic reservoirs of human diseases.

At about the same time, Matt Sullivan from Sallie Chisholm's lab at the Massachusetts Institute of Technology contacted me about sequencing a phage that infects *Prochlorococcus*. *Prochlorococcus* is a genus of cyanobacterium that carries out at least 25% of the photosynthesis on our planet. When we had completed the first cyanophage genomes, we were surprised to find a number of well-known photosynthesis genes. Had we messed up and sequenced host DNA? Or was it possible that these important genes are actually part of the phage genome? The answer was provided by Nick Mann and colleagues, who beat us in the race to publish evidence that the cyanophage genomes they had sequenced had a very special gene known as "*psbA*." It turned out that most marine cyanophage carry *psbA*, whose product allows photosynthesis to continue while the host is being killed by phage. Subsequent studies showed that many of the *psbA* genes in *Prochlorococcus* genomes have actually been through a phage intermediate.

By 2008, we had accumulated enough metagenomic data to determine the scale at which the global virome was contributing to the metabolic potential of their hosts. Compared to the microbes, the phage were enriched in genes with functions such as nucleic acid metabolism. More striking was the fact that the viromes also carry many genes associated with other unexpected aspects of metabolism (vitamin, cofactor, cell wall, and capsule synthesis), as well as genes for virulence factors, stress response genes, and chemotaxis genes. Thus, phage represent a vast, diverse repository of genes.

The phage-driven evolution of microbial hosts extends beyond the phage themselves. Working with Phil Hugenholtz and his group at the Joint Genome Institute, we were able to show that many of the changes in microbial metagenomes were actually responses to predation by phage and protists. This phenomenon seems to be very common. A most exciting recent development has been the discovery that host cells have rapidly evolving sequences, known as CRISPRs (for "clustered, regularly interspaced, short palindromic repeats"), that serve as a bacterial immune system to prevent phage infections in bacteria. Large-scale sequencing of phages and microbes from shared niches shows that there is a massive selection on microbes to avoid phage predators.

We now know that phage-microbial evolution is much more rapid and dynamic than Darwin could have imagined. Phage transport genes all over the world and change the evolutionary trajectory of the microbes they infect by introducing genes that completely change the phenotypes of their hosts. And by killing the microbes, the phage are forcing their hosts into a never-ending arms race. So most of the evolution on the planet is actually being carried out by entities Darwin never imagined and at a scale he never could have considered.

Microbes and Evolution: The World That Darwin Never Saw
Edited by R. Kolter and S. Maloy

doi:10.1128/9781555818470.ch10

10

The Struggle for Existence: Mutualism

Paul E. Turner

So it is said that if you know your enemies and know yourself, you will fight without danger in battles.

Sun Tzu, *The Art of War*, 6th century BC

If you can't beat them, join them.

Unknown

I am a microbiologist who loves the arts. I appreciate the emotional depth captured by a Diego Velazquez portrait, as well as the emotional lift of a perfectly catchy 1960s Motown hit single. Truly inspiring is the ability of a professional artist, songwriter, or author to maintain such creative highpoints over a long and illustrious career. Perhaps I am duly impressed because—although not widely acknowledged—the pursuit of scientific discovery is also a creative art form, as wonderfully espoused by W. I. B. Beveridge in his 1953 book *The Art of Scientific Investigation*. Successful scientists are not simply automatons, blindly adhering to dusty protocols and thoughtlessly incorporating shiny new technologies into their work. Rather, ground-breaking science is achieved by those who

Paul Turner obtained his Ph.D. at Michigan State University and did postdoctoral research at the University of Maryland College Park, University of Valencia, and the National Institutes of Health. Since 2001 he has been a faculty member at Yale University. Aside from constantly attempting to view the world from the perspective of viruses, he obsesses over collecting and listening to a wide variety of music.

rework established methods and ingeniously capitalize on technological advances, creatively synergizing existing and new approaches to unlock the mysteries of the natural world.

And what amazingly creative insights Charles Darwin possessed, revealed to us in his famous 1859 book, *On the Origin of Species*! For millennia humans had been modifying organisms to suit our whims, and it was plain to Darwin that such descent with modification could be achieved in agriculture because we purposefully chose which variants gave rise to future generations. Applying this knowledge to the struggle for existence that organisms experience in the face of natural challenges, Darwin wonderfully surmised that natural selection could account for Earth's biodiversity, giving rise to what Darwin described in his book as "endless forms most beautiful and most wonderful." Commemorating the 150th anniversary of his book's publication, in 2009 the Yale Center for British Art held the exhibit "Endless Forms: Charles Darwin, Natural Science and the Visual Arts," describing the profound and lasting influence that Darwin's scientific contributions have had on artistic views of the natural world. Indeed, it was a very fitting tribute to one of humankind's most creative scientific minds and an excellent example of how science and the arts often commingle.

In the early 1800s the Reverend Thomas Malthus produced essays on population growth, warning that humans were reproducing faster than our capacity to feed ourselves. With Malthus as his muse, young Darwin was inspired to ponder how the natural world was composed of finite resources, which caused a struggle for existence among variants within a population. Thus, with Nature as their muse, organisms evolve creative solutions to environmental challenges, and natural selection is the process that primarily dictates whether these challenges are successfully met through the evolution of adaptive traits. One apparent theme in the evolution of life is that natural selection often pushes organisms to take on parasitic lifestyles as their creative solution. If resources are often limiting and physiologically expensive to acquire or produce, an efficient solution is for organisms to thrive by capitalizing on the actions of others. For this reason, parasitism is often claimed as the most common species lifestyle on the planet; apparently in the biological world "there's a sucker born every minute" (a saying often credited to American showman P. T. Barnum) and even more organisms waiting in the wings to pounce upon their resources.

How often is parasitism the favored solution in the microbial world? Because we tend to selfishly focus on our own health and well-being, humans take special notice of bacterial pathogens which cause disease in humans and domesticated plants and animals. But these disease parasites are likely only a small fraction of the immense undescribed biodiversity in the microbial world. Doubtless, many of the seemingly benign bacteria in our world are only a handful of mutations or horizontal gene transfer events away from becoming opportunistic pathogens that can cause disease in humans. But the point is that we still know very little of the natural interactions occurring between bacteria; they may make up the majority of Earth's biomass, but we have few clues regarding the functional roles which most bacteria play in each other's lives and whether these interactions tend to be antagonistic or mutualistic. That said, consider the viruses which inhabit our planet. Although these residents of Earth are extremely small, they numerically dominate the biosphere, perhaps outnumbering bacteria 10 to 1. Viruses are noncellular and must infect metabolizing cellular hosts in order to undergo reproduction. Thus, viruses are obligate parasites, and their numerical superiority indicates that playing the role of the bad guy can be enormously fruitful in our world.

So what is a good guy to do? Fortunately, the righteous path of cooperation has not been forgotten by natural selection. Organisms may be pushed to lay down their arms and declare détente, by evolving to become mutualists that cooperatively support each other in the struggle for existence. In chapter 8 of *Origin of Species* Darwin described the remarkable interactions between ants and aphids; ants feed on the sugary honeydew excreted by aphids in exchange for protecting them from predators. Perhaps more remarkable is a three-party mutualism which also involves ants but, like much of the microbial world, was invisible to Darwin. Long before humans invented agriculture, leafcutter ants evolved to tend subterranean gardens of a tasty fungus. The fungus exists only in the underground chambers of the ant nest and is grown by cultivating the fungus on leaf material. This mutualism is augmented by another partnership with a species of *Actinobacteria* which lives in specialized glands on the underside of the ants and secretes an antimicrobial that kills other fungi attempting to invade the garden. Other amazing mutualisms involving microbes and multicellular organisms are awaiting discovery, which will similarly rewrite our understanding of the mutualisms documented by Darwin and other naturalists of his time.

Laboratory experiments with microbes also demonstrate that selection can create intimate partnerships, even when our artificially contrived environments would seem to prevent the formation of mutualisms. Like that of many other microbial evolutionary biologists, my work has involved selection experiments with the bacterial workhorse *Escherichia coli*. When a population of *E. coli* is grown in the laboratory on a single limiting resource such as the sugar glucose, it is expected that selection will favor whichever variant of the bacterium is the best grower, and through time the population will duly become dominated by this favored genotype. Thus, the population should be polymorphic only for the period when two or more variants are vying for dominance, and this periodic selection will continually occur until the population has exhausted the supply of spontaneous mutations which contribute to improvement in fitness on the sole limiting resource.

However, I and other researchers have observed a curious outcome whereby multiple variants of bacteria can evolve to coexist on the single resource, despite the classic prediction by Russian ecologist Georgii Gause that this should be impossible (Gause's Law, also known as the Competitive-Exclusion Principle). What we have learned from these observations is that bacteria and other microorganisms will not necessarily sit idly by while one variant evolves to become the best grower on the single available resources. Rather, one or more other variants in the population will evolve to feast upon the metabolites or other waste products of the efficient grower, seizing upon its efficiency by utilizing what the avid reproducer discards. Interestingly, such experiments sometimes suggest that a true mutualism may form, because selection can cause the better-growing microbe to also benefit by using some metabolite produced by its less efficient brethren. This give-and-take of resources is generally called cross-feeding, and the laboratory observations have spurred microbiologists to consider what may often be the course of evolution within natural communities of bacteria. For example, bacteria inhabiting the human gut beneficially aid our digestion of food, but it has been recently shown that these bacterial species may benefit one another through cross-feeding that helps ensure that they thrive despite unexpected changes in available resources in the human gut, owing to our omnivorous lifestyles.

The downside to bacterial mutualism is that it may teeter on a knife's edge depending on uncontrolled aspects of the environment. For example, perhaps certain species or variants of your gut bacteria would fare especially poorly the next time you decide to follow the latest fad diet. My colleagues and I discovered that a seemingly tried-and-true example of cross-feeding in *E. coli* did not last when the two partners were forcibly relocated from a laboratory in Michigan State University to the ivy halls of Yale University. Upon relocation, the more efficient grower simply refused to honor the presence of its slower-growing partner, and the established mutualism reverted back to a less interesting (for us) monoculture. It took much effort to finally realize that a breakdown in the polymorphism was due to an uncontrollable difference in the water used to create growth media in East Lansing, Michigan, versus New Haven, Connecticut. In particular, detectable differences in dissolved organic carbon between the water provided to the two labs probably accounted for the ability of a single *E. coli* variant to dominate in one location and of the two strains to cohabitate in another. This news is ominous for those researchers who seek to study and unravel the complexities of mutualisms which evolve in the wild. If we are challenged to study de novo evolved mutualisms between *E. coli* strains by moving them between labs, it suggests that we face a daunting task to recreate in the lab those environments which foster naturally evolved mutualisms between bacteria.

Why can't we all just get along? Let's leave aside obligate parasitic viruses for the moment—after all, viruses have a hard enough time convincing detractors that they are even alive, let alone representative of the planet's dominant lifestyle. Maybe for the most part, bacteria tend to foster each other's growth either directly or indirectly, and we have simply neglected to discover this triumph of mutualism over antagonism. Microbiologists are guilty of less often chasing down evidence for such mutualisms. We tend, rather, to focus overwhelmingly on the "bad" bacteria that do us harm and on the ability of microbes to produce antimicrobials that inhibit each other's growth (again, here we are strongly interested because we can co-opt these natural products to thwart disease pathogens). In defense of microbiologists, I would offer that such mutualisms are inherently difficult to study. We sometimes stumble across them in the laboratory when culturing bacteria on a single limiting resource, often because a stable polymorphism gives itself away, such as through a researcher detecting a difference in bacterial colony formation that persists through time. Now that we have several observations occurring across experiments with different microorganisms, the hope is that we can somehow use modified or novel technology to better probe for the existence of mutualisms, instead of bumping into them fortuitously. In addition, the growing interest among researchers in community ecology to determine whether community diversity begets community stability yields hope that microbiologists will similarly continue to appreciate whether beneficial community interactions between bacteria foster their long-term persistence in diverse assemblages in environments such as the human mouth.

In the wake of Darwin's bicentennial, we continue to laud his accomplishments, praise his legacy, and wish him many more happy birthday celebrations. My only pity for this pillar of science is that he missed the endless forms most beautiful and most wonderful in the world of microbiology, and how natural selection has shaped their many interactions. But then again, so far we have also missed the vast biodiversity and interactions hidden away from the naked eye, and my hope is that the creativity of scientific research during the next 200 years will increasingly bring them to light.

Microbes and Evolution: The World That Darwin Never Saw
Edited by R. Kolter and S. Maloy

doi:10.1128/9781555818470.ch11

11

The Secret Social Lives of Microorganisms

Kevin R. Foster

I remember asking myself, as I stood there in the sweltering attic of an English pub, "Why do I do this?" My face was a few centimeters from an unusually large hornet's nest, and things were about to get much worse. I tentatively put a bag around the brittle nest, but it fell from its moorings and exploded into a spray of hornets that immediately filled the attic. With a bee veil protecting me, I proceeded to collect the insects with my net, one by one. But growing impatient, I swung wildly and managed to shatter the bulb that hung from the center of the room. I was left standing in total darkness, covered in angry hornets.

The reason I was there is that hornets not only are strikingly beautiful creatures (when observed under better conditions) but also are an amazing product of social evolution. Like those of many bees and ants, their highly organized societies contain a single queen and a small army of specialized workers that ferociously defend and tend the young. What is remarkable is that many of these workers never reproduce. Any reading of Darwin immediately reveals a problem here: why would natural selection favor an organism that gives up its own reproduction?

Kevin Foster studied at the Universities of Cambridge and Sheffield in the UK as a zoologist. After fellowships at Rice and Harvard, he is now Professor of Evolutionary Biology at the University of Oxford. If not out chasing a bug of some description, he enjoys climbing up rock faces and hurtling down mountains.

Darwin himself saw this difficulty and devoted a section of *On the Origin of Species* to the problem of worker sterility and the associated issue of how, once sterile, an individual could be subject to natural selection at all. (If you do not reproduce, how can your traits be modified down generations into a specialized worker caste?) The answer, of course, is family life. Helping to raise your brothers and sisters—to whom you are genetically similar—is a perfectly good way to pass on copies of your genes. Familial gene sharing is so powerful that it can lead to the extreme case of the insect worker who labors her whole life without laying a single egg. The social insects have certainly earned the "social" in their name, and many biologists go even further, calling their selfless acts "altruistic" by analogy with human giving. But how far does this analogy go, and indeed can we take it further to the smallest of organisms? It was this question that led me, after my Ph.D., to move from insects to the world of microbes. I hope to convince you that, in important ways, microorganisms are altruistic too.

This idea seems ridiculous to some, and of course, there is no sense in which a microorganism ever displays the good intentions that often characterize human altruism. There is no cognitive root to their miniature gifts and exchanges. Nevertheless, the notion of the altruistic microorganism has a surprisingly long history, almost as long as the concept of natural selection. Herbert Spencer, a contemporary of Darwin, is widely credited with introducing and popularizing the term "altruism" in Victorian Britain alongside the term's probable inventor, Auguste Comte.

Spencer was a powerful intellectual figure who strongly promoted the importance of competition and evolution in human society through his catch phrase "survival of the fittest." Spencer used microbes in his discussions of social evolution and, moreover, suggested that there were important ways in which single cells were altruistic.

But I should not get ahead of myself. Before evaluating the idea of altruism, I will first consider the more general case for sociality in microbes. Is there any sense in which a bacterium has a social life? This idea too has been long aired. A particularly clear statement comes from Prince Pyotr Alexeyevich Kropotkin, a Russian anarchist who made several epic journeys into 19th-century Siberia. Amid the extreme cold and harsh conditions, he was struck by the importance of social interactions and, in particular, mutual aid among species. Kropotkin used these observations to bolster his argument for a decentralized and communist Russia. But it was his prediction that "we must be prepared to learn some day, from the students of microscopical pond-life, facts of unconscious mutual support, even from the life of microorganisms," that has had more lasting success.

Kropotkin's prediction, however, has taken time to reach the limelight. Modern microbiology rests upon the technique of vigorously shaking cells in liquid broth. This is ideal for many microbes: a washing-machine world where nutrients and oxygen are plentiful, a world where generations of microbiologists have studied the genetics and behavior of their favorite species. But the considerable benefit of growing cells in liquid can come at the cost of conceptual bias. The natural tendency for scientists is to picture species as single cells in a broth sea, where interaction is both rare and fleeting. This is at least partly correct. Many microbes, including the myriad that are found in marine plankton, do spend much of their lives swimming. One mode of swimming is the tail-like flagella of bacteria that spin rapidly and propel cells forward, a telling example of how biology invented both wheels and motors billions of years before we did. Flagella teach us that microbes are well adapted to a life in suspension, but there is another side to microbes that we were slower to recognize: a life in society.

We now know that much of what microbes do they do in dense groups of millions. Many bacteria like to stick themselves to surfaces where they can readily proliferate and, while doing so, encase themselves in sticky or slimy substances. These gatherings, known as biofilms, are found in all manner of environments. One such environment is your teeth, where millions of cells of many species are currently growing as incipient plaque. Biofilms form on almost any surface, including those deep inside our bodies and upon medical devices, ship hulls, and food-processing equipment. The problem is compounded by the fact that biofilms can be tough and confer considerable antibiotic resistance on their inhabitants, making them difficult to remove. On the flip side, some such communities are vital for our health and well-being, including the microbial communities that clean our water in treatment plants and the massive numbers of bacteria that line our gut. An understandably favorite factoid of the microbiologist is that the bacteria within us outnumber our own cells by 10 to 1.

Everything that a cell does within a biofilm can have immediate consequences for its neighbors. This includes the most fundamental action of any microbe: cell division. Not only does division push others aside, but also it requires many resources that can be scarce in a densely packed group. The act of cell division is thus the act of taking food from others. Accordingly—as for a rebel reproducing worker in a hornet's nest—natural selection can favor cells that divide as rapidly as possible in order to acquire as great a share of the resources as possible. Recent events in both the economy and the environment reveal the human analogies to this situation. If left unchecked and unregulated, competition within any society has the potential to waste—and eventually ruin—shared resources.

This idea was summarized in a famous essay by the ecologist Garrett Hardin. Hardin built his argument upon a story of a commons pasture in rural England shared by multiple herders. He observed that the best way for a herder to maximize productivity is to keep adding cattle to graze, even though this threatens the pasture's demise. This curious situation arises because with a shared resource, all the benefit of adding a cow goes to the owner, while the cost to the pasture is shared among all herdsmen. From here, Hardin argued strongly—and controversially—that we need to find ways to regulate human population growth in order to avert environmental tragedy, an argument that may be more likely to be heeded now than it was originally.

The potential for Hardin's "tragedy of the commons" is a problem shared by microbes and humans alike. Natural selection can favor microbes that divide as rapidly as possible in order to maximize their short-term gain, irrespective of the negative effects on others. Conversely, if a cell holds back on division and increases the total yield of a biofilm, the cell is displaying behavior that will have a positive effect on others. In this simple sense, a microbe can be said to be altruistic. Cells have the potential to display behaviors that slow their own reproduction and promote that of others. But why would a cell ever do this? Here, as for social insects, the key is again likely to be family life. As bacteria divide in a packed biofilm, they produce near-identical copies of themselves. This can create large clonal groups with a common evolutionary interest, even if the biofilm as a whole contains many competing strains and species. Consistent with the notion of altruistic behavior within biofilms, many microbial species are now known to secrete products that are costly in the short term but improve overall growth. These products include enzymes that break down food sources to a manageable size and scavenging molecules that help collect scarce resources such as iron. Because the products are released into the environment, they are a shared resource and the secretions of any one cell will benefit many others around it.

Altruism is beneficial only if there are other cells around to receive the benefits. If a cell were to sit alone and secrete a growth-promoting compound, there would be little benefit to anyone, as most of the compound would diffuse away and be lost. Microbes have evolved to deal with this problem by secreting cheaper compounds that are used to detect cell density (and possibly diffusion conditions). When these indicators reach a high enough concentration, the cells can sense that there must be many other cells around, and the conditions are ripe for secreting the more costly products.

This clever trick is, of course, quorum sensing. An effective society of microbes thus requires several of the key features of any prosperous society. This includes the need for individuals to perform selfless acts and the need for coordination that ensures that these acts are well directed. The analogies to other societies are also illustrated by the potential for mutants that use the secretions of others without themselves contributing to the shared pool. This "microbial cheating" occurs both at the level of quorum sensing—where one strain of bacteria uses a quorum sensing signal without producing it—and at the level of the many secretion systems that provide nutrients for growing cells. The emergence of such strains tends to decrease the growth rate of microbial groups.

What can be very bad for a biofilm, however, can be very good for us. Many pathogens rely on both quorum sensing and secreted products to inflict harm upon us. This includes the secretion of enzymes that kill host tissue by *Salmonella enterica* (typhoid fever) and *Pseudomonas aeruginosa*; toxin production by *Bacillus anthracis* (anthrax) and *Vibrio cholerae*; and the widespread production of compounds that break down antibiotics, such as the β-lactamases that destroy penicillin.

Or is there another way? Rather than focusing on antibiotic use within our societies, we might instead focus on microbial societies and the potential for their own microtragedies (Fig. 1). Traditional antibiotics act by killing or stopping cell division, and resistant

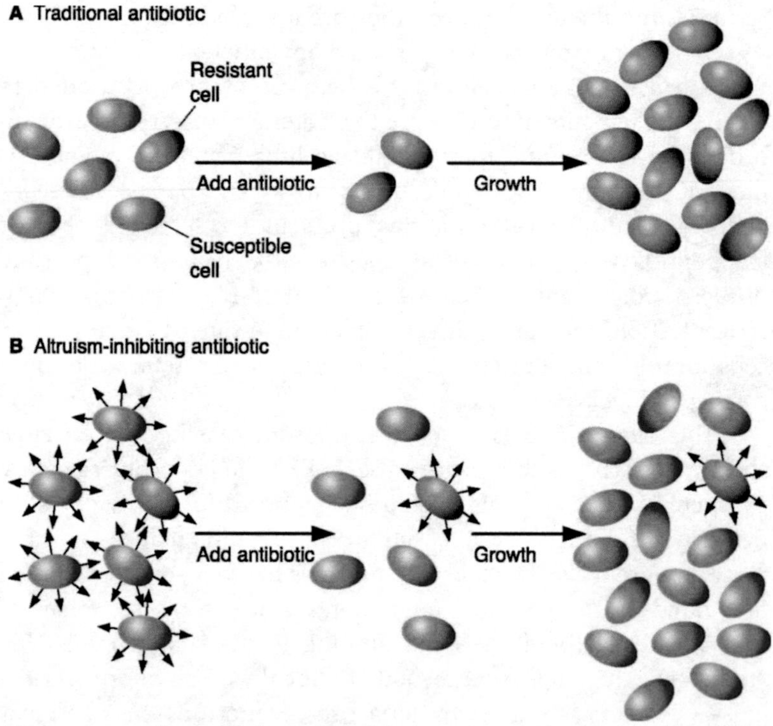

Figure 1 The evolution of antibiotic resistance in a typical antibiotic (A) and a hypothetical antibiotic that targets a secretion, where the secretion is energetically costly to individual cells but promotes growth of nearby cells (B). In the conventional case, the resistant mutants rapidly outcompete the susceptible cells. In the latter case, resistant mutants that secrete have the potential to be outcompeted by susceptible cells that do not, because the susceptible cells can use the growth-promoting secretion without paying the cost. doi:10.1128/9781555818470.ch11f1

mutants that can grow in the presence of the antibiotic rapidly replace the original susceptible strains. Consider instead a drug that attacks a cell's ability to secrete an altruistic compound that is needed for growth. Resistant mutants that re-evolve secretion will promote the growth of susceptible cells around them. And, more than this, the susceptible cells do not pay the cost of secretion, which can put the resistant strain at a competitive disadvantage. At least in principle, this can slow the rise of antibiotic resistance. By recognizing that microbes rely on both sociality and altruism to cause infection, a novel strategy for treatment is revealed.

FURTHER READING

Andre J, Godelle B. 2005. Multicellular organization in bacteria as a target for drug therapy. *Ecol Lett* **8:**800–810.

Foster KR, Grundmann H. 2006. Do we need to put society first? The potential for tragedy in antimicrobial resistance. *PLoS Med* **3**(2)**:**e29.

Kolter R, Greenberg EP. 2006. Microbial sciences: the superficial life of microbes. *Nature* **441:**300–302.

Kropotkin PA. 1902. *Mutual Aid: A Factor of Evolution.* McClure Philips & Co, New York, NY.

Nadell CD, Xavier J, Foster KR. 2009. The sociobiology of biofilms. *FEMS Microbiol Rev* **33:**206–224.

Microbes and Evolution: The World That Darwin Never Saw
Edited by R. Kolter and S. Maloy

doi:10.1128/9781555818470.ch12

12

Microbes and Microevolution

Evgeni Sokurenko

I have a huge black Labrador retriever. Like any Lab, he loves to swim and greets every stranger as his favorite-in-whole-world aunt or uncle. The only things he loves to maul are huge logs pulled out from the water. But my good neighbor has a pit bull. She hates water and, though she has known me for many years as a friend of her master, always greets me with a horrifying growl. (And I do not want to go into what pit bulls would love to maul.) So, Labs behave quite differently from pit bulls. Shape-wise, what can be more different than Chihuahuas and Great Danes? And then there are gray wolves, which are...wolves. The amazing part is that all this variety of animals belongs to the same biological species—*Canis lupus*. With the domesticated dogs, it is selection by humans that produced the most incredible within-species variety of animals known to us. For thousands of years, we were picking on small and random variations noted in the exterior appearance or behavior of dogs that could be fixed by breeding and eventually produce the desired, and often huge, differences seen today among different breeds. So, these within-species differences are inheritable, i.e., are based on genetic mutations that are passed from one generation to another.

Evgeni Sokurenko obtained his MD and Ph.D. at the Moscow Medical (Sechenov) Institute, USSR. He joined the microbiology faculty at the University of Washington, Seattle, in 1999. In the little time left from writing research papers and grants, he enjoys running, weightlifting, kebab-grilling, and arguing with his friends about the quality of different brands of vodka.

Humans used the same artificial breeding techniques to produce the great within-species variety of farm animals, with desired and distinct properties. And from this alone, one can see how much Darwin was right by pointing out the ability of notable inheritable variations to emerge within same species. Emergence of such within-species differences in the size and shape (morphology) or behavior of individual organisms has been called microevolution. What Darwin also postulated is that these random, within-species variations can be subjected to natural selection to the extent that they would result in novel species that stop interbreeding with the original ones—the phenomenon called macroevolution. Opponents of evolution cannot deny the obvious successes of human-driven selection of domesticated animals and generally admit the processes of microevolution in nature. What they question is whether natural selection (i.e., not based on an intelligent choice by humans) can result in macroevolution and produce the variety of different species existing today, with a huge range in morphological and behavioral complexity.

I am a true believer in Darwin's theory that the existing complexity and variety of the natural world evolved naturally. However, it is not the main intention here to argue for the reality of macroevolution. Rather, the main focus is on a much less controversial topic: microevolution. At the same time, what I would like to show is that minor genetic changes can lead to a dramatic shift from the original natural environment (habitat) of the species to completely new ones. In turn, these shifts in habitats could lead to physical separation of the organisms from the same species and their eventual macroevolution into different species.

So, how much could be the natural variability within a given species, and how much does it matter for their ability to thrive in distinct separate environments? One of the best examples of microevolution can be found among microorganisms, in particular bacterial species able to cause infections in humans, i.e., pathogens. The variability in the ability to cause disease can be astounding even for organisms in the same species. Let's take the most famous bacterial species, *Escherichia coli*. *E. coli* can be isolated from the feces of the vast majority of healthy humans and is mostly an inhabitant of the large intestine, where it obviously does not do any harm. Moreover, to stay healthy, some recommend taking Muta-Flor capsules regularly, which contain billions of live bacteria of *E. coli* strain Nissle-1917. However, *E. coli* can also cause a great variety of some of the most common bacterial diseases in humans: diarrhea, sepsis, and urinary tract infections, to name a few. And consuming just 10 bacteria of some *Shigella* species, which is a relative of *E. coli*, can kill us. Most other bacterial pathogens also belong to bacterial species with a very wide variability in the ability to either cause disease or be generally harmless.

To discuss the basis and significance of natural mutations within bacterial species, it is actually useful to make parallels not with domesticated animals but with ourselves, humans. On one hand, we cannot be further apart from bacteria. Humans are built from a trillion cells, physiologically and behaviorally one of the most complex organisms, compared to unicellular bacteria, which are 10 times smaller than the smallest human cell, lack a nucleus, and usually have a single small chromosome. Also, as much as humans can look (and act!) differently, these differences among us do not amount to anything really drastic, especially compared to our pets. However, two features unite humans closely with bacteria: we both are tenacious explorers of new habitats around us and are able to adapt quickly to different environments.

The ability of bacteria to spread around lies in their small size. They can do it with flowing waters and blowing winds and on the feet of wild animals, the wings of birds, or the hands and lips of human beings. When they encounter humans, bacteria can enter even the tiniest pores of the body and can actually use human cells as vehicles to move around. This ability of the bacteria to spread prompted Dutch microbiologist Lourens Baas-Becking to advance in 1934 the famous hypothesis regarding the microbial species that "Everything is everywhere, but the environment selects."

The second part of the Baas-Becking hypothesis is related to the other important similarity between humans and bacteria, i.e., adaptability. We humans adapt to different environments not by changing genetically but by inventing various tools that allow us to successfully survive and increase in numbers in new environments. Using the plow and sickle helped to move farming deep into continental Asia and Europe, warm boots and clothes led us further into the cold areas, better spears and fishing nets improved the chances of survival in areas with scarcer resources, military inventions led some groups to purge others from valuable areas or protect from invaders, etc. Furthermore, what also helped the human species to progress is the specialization of labor, with farmers getting better in farming, hunters in hunting, and soldiers in fighting. Like all other living creatures, bacteria, of course, do not adapt through inventing material tools; they use genetics instead. But they are very good at that. There are two major mechanisms through which organisms from the same bacterial species can adapt: so-called horizontal gene transfer and mutations.

Horizontal gene transfer is the acquisition of novel genes from other bacteria. In the analogy of humans, this is reminiscent of getting ready-to-use tools instead of making them from scratch. For example, if life forces a peaceful farmer to become a thug, he could obtain (buy or steal) a dagger specifically designed to be used as a weapon. Indeed, horizontal gene transfer is a very common mechanism of adaptation for bacteria specifically, as such transfer is not commonly found in nonbacterial species. Such a mechanism involves acquisition of individual or multiple genes from bacterial viruses (bacteriophages), small self-reproducing circular DNA plasmids, or large multigene insertions into the chromosome, so-called chromosomal islands. Horizontal gene transfer accounts to a great extent for the within-species variability of bacteria and, in many cases, could be the critical difference between harmful and harmless organisms. Indeed, in pathogens, these extra genes can provide the means to invade human cells, protect from the immune defenses, attach to tissues, or induce human cell damage through secreted toxins.

Despite the great importance of horizontal gene transfer for bacterial microevolution, it does not answer the key question: how did genes with novel function originally evolve? To answer this, we need to discuss the second mechanism of bacterial adaptation, i.e., mutations. Mutations are changes in the existing genes. Some of them can be large and result in either duplication or deletion of one or multiple genes at a time. Most of the mutations, however, are relatively small and often affect just a single nucleotide. Unlike horizontal gene transfer, which is common mostly for bacteria, mutations occur in all organisms, from viruses to humans. Mutations are the main source of novel variations and are the primary force behind both micro- and macroevolution. For example, the evolution of humans from apes is likely to involve many mutational changes selected over time. Mutational changes also accumulate within species, and while many of them can be neutral in nature, others could provide relative advantages or disadvantages for organisms. In humans, for example, most attention is drawn now to finding mutational changes that result in inherited disorders or the predisposition of some individuals to infections or other illnesses, like stroke or heart attack. In contrast, in bacteria, specifically bacterial pathogens, most of the attention is given to mutational changes that are in one way or another advantageous for them. While, unfortunately for us, mutations of this type make pathogens more resilient to our body defenses or therapy (including antibiotics), they represent some great examples of how small genetic changes can be adaptive for novel environments.

As much as adaptive mutations seem to be important for microevolution, very few such natural changes are found within species of higher organisms. They accumulate relatively slowly, and it could take many generations for the advantageous effect to become prominent in the natural population. In contrast, because bacteria multiply fast (some divide every 10 minutes), they provide a great opportunity for such mutations to arise and be detected in real time. In pathogens, such mutations are called pathoadaptive and can be acquired in the course of an infection in a single patient. Such mutations can make bacteria able to produce, in much larger amounts, some toxin or material that protects them from the body's defense or an antibiotic's action. They can also modify the structure of bacterial proteins that originally served as tools of a harmless lifestyle into ones that can be effectively used to increase their success as pathogens. This would be analogous to the farmer-turned-thug not going to the trouble of obtaining a dagger but converting his kitchen knife into an effective weapon by making it, for example, double edged, or making a spear from a shovel.

Such mutations in pathogens could be very effective. For example, *Pseudomonas aeruginosa*, a bacterium that inhabits water streams around us, can cause very dangerous lung infections in individuals with cystic fibrosis, a condition in which the lung mucus becomes very thick and hard to clear up. In the course of such an infection, which is chronic and lasts for many years, the infecting bacteria acquire a mutation that knocks off a regulator of the production of surface capsular material, making them overproduce it. This overproduction protects the bacteria very well from immune cell attack or antibiotics, making it impossible to get rid of the lung infection and leading eventually to the patient's death. Another example is single changes in the structure of the major cell attachment protein of *E. coli* strains that cause infection of the urinary bladder and kidneys, one of the most common and potentially debilitating bacterial infections in humans. These mutations make the protein (which is originally designed to ensure harmless bacterial colonization of intestine) adhere 10 times better to the urinary tract cells, making it much harder to remove bacteria from the normally sterile urinary compartments. Yet another interesting example are point mutations in one of the secreted enzymes of soil inhabitant *Listeria monocytogenes*; these make the bacterium capable of entering and surviving inside human defense cells (macrophages) and as a result cause one of the most severe food-borne infections of the bloodstream, in which up to a third of cases result in death.

These examples of adaptive mutations illustrate how significant shifts in the inhabiting environment of bacteria—from water stream to lungs, large intestine to urinary tract, soil to bloodstream—can result from single or very few changes in the DNA. The successful invasion of novel environments provides an opportunity for the bacteria to continuously adapt there. This can lead to accumulation of additional differences between the originally harmless and the evolved pathogenic bacteria, and can potentially lead to emergence of completely separate species, adapted exclusively as human pathogens; for example, tuberculosis-causing *Mycobacterium tuberculosis* emerged from a cow-adapted species much less harmful to humans. In other words, within-species microevolution with the habitat shift provides a basis for between-species macroevolution.

Another illustration provided by adaptive mutations in bacteria is that a novel function can be acquired in a gene (or coded protein) by minor structural changes. On one hand, analysis of the functional effect of the natural mutations selected in the genes (proteins) allows us to better understand how they actually work. Thus, nature serves as a huge research laboratory for us; we need only to figure out how to detect and interpret the results natural selection came with. On the other hand, the shifts in the function open the possibility for the gene to evolve further in a certain direction, step by step, leading to more and more structural changes and functional specialization that deviate more and more from the original. For example, some bacterial toxins that are used as a weapon against immune cells could evolve from bacterial virus proteins originally adapted to lyse bacteria themselves. Considering the tremendous population size of bacterial species (e.g., there are 10 to 100 times more bacteria inhabiting our intestine than there are cells in our body), mutational changes can affect not only all possible genes but also all possible nucleotides in the DNA within a single population of bacterial species. Combined with some efficient mechanisms of gene shuffling between bacteria, this provides an opportunity for a buildup of new networks of genes with new functions, eventually leading to the emergence of novel complex metabolic pathways or such fascinating multiprotein structures as the flagellum, a locomotion organelle of bacteria in which individual components seem to be perfectly coadapted for a single function.

Thus, studying microevolution of bacterial species can bring light to the elusive processes of microevolution in higher species. As the Nobel Prize winner Jacques Monod once said, "What is true for *E. coli* is true for the elephant." In turn, an understanding of the microevolution mechanisms and dynamics will bring us closer to grasping how new species and complex physiological systems have gradually evolved. Indeed, it took a long series of small and big steps to progress from invention of the wheel to manufacturing the latest Lexus models. After all, natural selection of adaptive mutations is as natural as selection of adaptive ideas—it just takes longer. Let's be patient, careful, but persistent in our quest for knowledge, and great insights will keep coming, like the one that came 150 years ago from Charles Robert Darwin.

Microbes and Evolution: The World That Darwin Never Saw
Edited by R. Kolter and S. Maloy
©2012 ASM Press, Washington, DC
doi:10.1128/9781555818470.ch13

13

Unnecessary Baggage

Stanley Maloy and Guido Mora

> I believe that disuse ... has led in successive generations to the gradual reduction of various organs, until they have become rudimentary,—as in the case of the eyes of animals inhabiting dark caverns, and of the wings of birds inhabiting oceanic islands, which have seldom been forced to take flight, and have ultimately lost the power of flying. Again, an organ useful under certain conditions, might become injurious under others,... and in this case natural selection would continue slowly to reduce the organ, until it was rendered harmless and rudimentary.
>
> Charles Darwin, *On the Origin of Species*

Humans are so adapted to thinking that everything has a reason that it is very difficult for us to grasp the probability of rare, random events. Even scientists succumb to this thought pattern, creating "just so" stories that would make Rudyard Kipling proud.

Stanley Maloy obtained his Ph.D. at the University of California at Irvine, then did postdoctoral research at the University of Utah. He was on the faculty at the University of Illinois in Urbana-Champaign for many years, and moved to San Diego State University in 2002. His favorite activity is thinking about science, but he also enjoys traveling and soaking his head in the ocean. *Guido Mora* did his academic training at Universidad de Chile and obtained his Ph.D. at the University of Wisconsin, Madison. He was on the faculty at the Pontificia Universidad Católica de Chile for many years, and moved to Universidad Andrés Bello in 2003, where he was Dean of the School of Ecology & Natural Resources. His favorite activity is traveling, and he is very grateful to Science for, among other things, giving him the chance to visit many different countries.

However, evolution occurs by rare changes that are simply due to random chance.

Although Darwin didn't know that genetic changes were the basis of variation between organisms, he clearly argued that evolution occurs by natural selection acting on preexisting random variations that are transmitted to offspring. We now know that this variation is due to spontaneous mutations. Sometimes mutations are beneficial and will increase in abundance because they help the organism reproduce more efficiently. Much more often, mutations are harmful and are lost because the mutant organism reproduces less efficiently than its nonmutant siblings. But, sometimes mutations aren't beneficial or harmful; they are just neutral—that is, they do not have any detectable effect on reproduction of the organism. For example, if a certain gene product is not needed in a particular environment, mutation in that gene will not affect growth in that environment. Hence, if a random mutation occurs in that gene, there is no selection for or against the mutant in the population. Nevertheless, these changes may be important when the organism adapts to a new environment.

Animals and plants have many genes that are defective because of such neutral mutations. Mutated genes that cannot make a functional product are called pseudogenes. When we were students, we were taught that pseudogenes must be extremely rare in bacteria compared to plants and animals because bacteria grow rapidly and have small genomes. Hence, the energy wasted by carrying nonfunctional genes would select against bacteria with them. At the time, most of our understanding of bacteria was based upon a few fast-growing strains of bacteria that could be studied genetically. For example, the majority of the detailed genetic studies that taught us about mutations were based upon a common laboratory strain of *Escherichia coli* named K-12 and a laboratory strain of *Salmonella enterica* named Typhimurium LT2.

One of us, Stanley Maloy, had worked for many years on the genetics of *Salmonella* Typhimurium LT2, dissecting the regulation of a variety of metabolic processes and developing new genetic tools to understand these processes. The other, Guido Mora, worked on a closely related *Salmonella enterica* serovar called Typhi, which causes a life-threatening disease in humans, typhoid fever.

In contrast to Typhimurium, which can infect many organisms from mice to humans, Typhi infects only humans. Not surprisingly, we didn't know each other. Our labs were in different hemispheres, and we worked on different topics. That changed at a scientific meeting in 1989 where Stanley was presenting his work on genetic tricks for studying membrane transporters and Guido was presenting his work on the biochemistry of outer membrane proteins. There were very few genetic tools for studying Typhi, so it was much less well understood than Typhimurium. Talking about our different research projects led to the suggestion that together we could develop tools to study this human pathogen, leading to a long collaboration between our two labs. In particular, we wanted to understand why Typhi infects only humans, while the very closely related Typhimurium is such a promiscuous pathogen. At the time, it was thought that most important differences between closely related bacteria were due to the acquisition of new genes from other bacteria, a process called horizontal gene transfer (as opposed to inheritance between generations, called vertical gene transfer). We worked together to develop some clever genetic tricks for identifying such "host specificity" genes in Typhi. After years of hard work by many students in our labs, it seemed clear that the reason why Typhi was restricted to humans was not because of the acquisition of new genes but because Typhi had mutations in many different genes scattered around the chromosome. However, it wasn't clear why these mutations restricted Typhi to humans. The insights into this question came from a completely different approach.

Often a new tool allows us to see things that we could not see previously. In the mid-1990s it became possible to determine the entire sequence of bacterial genomes. The genome sequences of different types of bacteria provided surprising insights into genes we had known nothing about previously. But the most striking insights came from comparisons of the genome sequences of closely related bacteria. For example, although many bacteria, like *E. coli* K-12 and *Salmonella* Typhimurium LT2, have relatively few pseudogenes, large numbers of them were found in bacteria that are obligate intracellular parasites and in pathogens that infect only a single host. These results indicated that the dogma that pseudogenes are extremely rare in bacteria was wrong. It is always exciting for scientists when a dogma is overturned, because this means that there is some previously unrevealed aspect of nature that is waiting to be uncovered.

In contrast to *Salmonella* Typhimurium, the human-specific pathogen *Salmonella* Typhi was one of these organisms with a large number of pseudogenes. In addition, many of the pseudogenes were in parts of the genome we had identified as containing mutations that restrict host specificity. We wanted to know whether the pseudogenes are responsible for restricting Typhi to the human host or whether the accumulation of pseudogenes is simply an accident. To answer this question, our two labs used slightly different approaches. Stanley's lab moved pseudogenes from Typhi into Typhimurium and determined whether the Typhimurium organisms with the Typhi pseudogenes lost the ability to infect mice. The results indicated that while some pseudogenes have no effect on Typhimurium infections, other pseudogenes cause a small decrease in the ability to infect mice. However, these effects were synergistic—that is, two pseudogenes that each have a small effect on infection in mice had a much greater effect on ability to infect mice when combined in the same bacterium. Guido's lab did the opposite, repairing the pseudogenes in Typhi with a copy of the corresponding gene from Typhimurium. The results showed that one of the pseudogenes from Typhi enhances important steps of the infection in human cells grown in the laboratory. Thus, the genome sequencing studies clearly indicated that pseudogenes occur in bacteria, and our results showed that pseudogenes can influence the host range of bacterial pathogens.

This raises the question of "the chicken and the egg": did Typhi become restricted to humans because it acquired pseudogenes, or did pseudogenes arise because when Typhi was restricted to humans the intact genes were no longer needed? Some of these genes that are required for infections in mice are not required for human infections—that is, the genes are neutral for human infections. As Typhi coevolved with a human host, these genes were simply excess baggage, but once these genes were lost, the bacteria became less effective at infecting other hosts. Some pseudogenes both enhance the infection of human cells and decrease the ability to cause infections in mice, further restricting Typhi to humans. This accumulation of mutations that further restricts the host range of Typhi is an example of a genetic process called "Mueller's ratchet," because it acts like a socket wrench that readily turns in one direction but not the other—each additional pseudogene adds another constraint on the ability to survive in the "old" environment.

So, although we expected to find novel genes that restrict *Salmonella* Typhi to human hosts or "murium" genes that allow *Salmonella* Typhimurium to infect mice, the results reaffirm the roles of chance and necessity in evolution: random mutations may accumulate in one environment because the gene products are not required in that particular niche, but accumulation of such mutations decreases the versatility of the organism and ultimately restricts the mutant organism to that particular niche.

FURTHER READING

Deng W, Liou S, Plunkett G III, Mayhew G, Rose D, Burland V, Kodoyianni V, Schwartz D, Blattner F. 2003. Comparative genomics of *Salmonella enterica* serovar Typhi strains TY2 and CT18. *J Bacteriol* **185:**2330–2337.

House D, Bishop A, Parry C, Dougan G, Wain J. 2001. Typhoid fever: pathogenesis and disease. *Curr Opin Infect Dis* **14:**573–578.

Matthews TD, Maloy S. 2010. Genome rearrangements in *Salmonella,* p 41–48. *In* Fratamico P, Liu Y, Kathariou S (ed.), *Genomes of Foodborne and Waterborne Pathogens*. ASM Press, Washington, DC.

Microbes and Evolution: The World That Darwin Never Saw
Edited by R. Kolter and S. Maloy

doi:10.1128/9781555818470.ch14

14

Bacterial Adaptation
Built-In Responses and Random Variations

Josep Casadesús

Computer scientists use the term "robust" for software that performs well not only under ordinary conditions but also in stressful, unforeseen situations. One can hardly think of biological systems more robust than bacteria. As pointed out by Gould, bacteria are the oldest forms of life and their lifestyle remains the most successful. Bacteria thrive everywhere on Earth, including extreme environments like the Dry Valleys of Antarctica and the geysers of Yellowstone National Park. The human body itself is the home for thousands of bacterial species, and the total number of bacterial cells in a healthy individual may be similar to, or even exceed, the total number of "human" cells.

Bacteria live in a changing environment, devoid of the homeostatic mechanisms that create stable conditions in the tissues of multicellular eukaryotes. Except for obligate parasites that have adapted to stable environments, survival of bacteria depends on ceaseless adaptation. The latter is possible because bacteria are equipped with biochemical software responsive to environmental changes: signal transduction systems, hierarchical gene networks,

Josep Casadesús received training as a bacterial geneticist in Granada, Salt Lake City, and Basel. He teaches Genetics at the University of Seville. Discussing science is one of his favorite activities. He also enjoys hiking, traveling, cooking, wine tasting, and listening to music in the early morning.

stress responses, DNA repair systems, efflux pumps, and so on. A combination of flexibility, graduality, and redundancy contributes to the robustness of the bacterial adaptive software. Furthermore, accidental differences in software usage by individual cells create epigenetic polymorphism that preadapts bacterial populations to environmental change. However, built-in adaptive capacity and epigenetic polymorphism may be insufficient to explain the evolutionary success of bacteria. Below I discuss how two concepts elaborated in the last decades have changed our view of bacterial evolution. One is the evidence that horizontal gene transfer is a major innovative force in the bacterial world. An additional, more controversial notion is that bacterial mutation rates can vary in response to environmental challenges.

Graduality and Overlap of Bacterial Built-In Responses

Many bacterial regulatory circuits are not digital but analogue devices. In other words, their responses are not all-or-none but are proportional to the stimulus to which they respond. For instance, attenuation adjusts the transcription rates of biosynthetic operons (tryptophan, histidine, and others) to the availability of charged tRNAs. Regulation of mRNA translation and/or mRNA stability by small regulatory RNAs is likewise gradual. Analogue circuits are obviously energy-wise, since their responses are adjusted to the actual need and waste efforts are prevented. Examples of graduality are found not only in transcription and translation of individual mRNAs but also in the expression patterns of gene networks. A well-known example involves the SOS regulon, which comprises a large number of genes under the control of a single repressor, LexA. When DNA damage occurs, LexA is degraded and SOS genes are turned on. The number of LexA molecules degraded depends on the magnitude of DNA damage. As a consequence, genes of the SOS regulon are turned on in an orderly, hierarchical manner depending on the affinity of the LexA repressor for their operators or "SOS boxes." Gradual SOS induction thus adjusts the response to the degree of DNA damage. For instance, the *sulA* and *umuDC* genes are among the last to be turned on. This permits the cell division inhibitor SulA to be synthesized only if DNA damage is massive and cell division arrest becomes necessary. In a similar way, error-prone DNA polymerase UmuDC is synthesized only under conditions that require DNA replication at any cost.

Overlap and redundancy also contribute to bacterial robustness. Examples of overlapping responses are very common in bacterial physiology. For instance, *Salmonella enterica* is exposed to bile in the gut of mammals. Exposure to bile activates the Mar regulon, a set of >60 genes involved in regulation of efflux pumps and other protective devices, as well as the SoxRS regulon, which contains genes for oxidative damage defense, and the RpoS-dependent general stress response. Evidence obtained in my laboratory suggests that activation of any of these individual gene networks might be sufficient to cope with the nasty effects of bile salts. However, simultaneous activation of several protective responses may increase robustness. Repair of DNA lesions caused by bile salts provides another example of overlapping, redundant responses. Repair of bile-induced DNA damage can be carried out by several pathways, including mismatch repair, base excision repair, recombinational repair, and SOS-associated translesion DNA replication. When *S. enterica* is exposed to bile salts, all available DNA repair systems, including the SOS response, deal with DNA lesions. However, activation of the SOS regulon in the presence of bile salts is only transient. A tentative interpretation is that the initial reaction of the bacterial cell is activation of every resource that can help. However, a potentially dangerous response like SOS induction is turned off as soon as other DNA repair systems are working at full speed.

Redundancy is also found in one of the most critical processes in bacterial life, DNA replication. To illustrate this point, let me tell you of an interesting discussion that took place years ago at a DNA replication workshop. With a mixture of wisdom and arrogance not rare among scholars, an expert on the eukaryotic cell cycle blamed bacteriologists for "not having solved" the bacterial cell cycle. The actual fact, however, is that the orderly series of events and checkpoints typical of the eukaryotic cell cycle does not exist in bacteria. The bacterial cell cycle is extremely flexible and can adapt its parameters to changing circumstances. Even the initiation of chromosome replication can occur in more than one way. For instance, in the absence of elements necessary for standard initiation, recombination can trigger the so-called "stable DNA replication" which does not require initiation at the *oriC*. As reasoned above for bile resistance, bacteria can often solve a problem using various solutions. The ability to face a challenge in several ways may be crucial for robustness.

Heterogeneity in Bacterial Populations

Built-in responses, no matter how finely tuned by natural selection, are probably insufficient to explain the adaptive capacity of bacteria. Bacterial populations also benefit from nondeterministic, random variations in their molecular circuitry—called epigenetic variation because the processes are heritable but not due to mutational changes in the DNA. Such epigenetic variation preadapts bacterial populations against sudden, unexpected changes. This is possible because all cells, including bacteria, undergo a substantial degree of noise at the molecular level, in the sense that their responses show a certain degree of random fluctuation. Far from being a problem, molecular noise can improve the ability of bacterial populations to respond to environmental challenges. For instance, a population of genetically identical bacteria can express certain genes in a nonuniform manner, thus generating two or more subpopulations. Epigenetic generation of two alternative cell states, a phenomenon known as bistability, appears to be relatively common in bacterial populations, as judged by single-cell analysis of gene expression. As an example, consider a protein that activates transcription of its own structural gene: the cell needs the protein to make more of it; if there is none, the cell cannot make any. Hence, the two steady states of this bistable system are "protein present, gene on" and "protein absent, gene off." Switching between bistable states can be caused by genetic rearrangements and by epigenetic changes such as the formation of DNA methylation patterns, thus generating phenotypic diversity without the permanent scars of mutation. However, epigenetic switching can also occur by random fluctuation, followed by positive or doubly negative feedback loops in the regulatory network. Whatever its origin, bistability (or, in a more general manner, multistability) is a source of heterogeneity, thus allowing bacterial populations to get ready for future challenges should they occur.

Bacterial Genome Plasticity

Extrachromosomal DNA has been traditionally viewed as the main toolbox for bacterial genome plasticity. Plasmids are crucial indeed for bacterial adaptation. However, the bacterial toolbox is not restricted to plasmids and includes also the chromosome. Before the advent of full genome sequencing, bacterial species were considered to have fairly constant chromosomes, and chromosome plasticity was thought to be mostly restricted to transposons and prophages. For instance, I used to tell my students that *Escherichia* and *Salmonella* had very similar chromosomes, except for a large inversion and a few genus-specific DNA blocks. In a similar way, *Shigella* was viewed as an *Escherichia coli* organism that had acquired a virulence plasmid and had gained and lost a few chromosomal genes. Such views were not just simplistic; they included the additional mistake of considering bacterial species more homogeneous than they actually are. For instance, a recent study undertook entire sequencing of 20 *E. coli* strains. The authors discovered that the strains shared a core genome of more or less 2,000 genes. However, the average genome size was 4,700 genes, and the total number of genes found in the *E. coli* collection was over 17,000. Hence, the strains under study could be called *E. coli* because they more or less shared a core genome. However, no single strain could be considered representative of the species. Aside from challenging classical bacterial taxonomy, the study confirms that horizontal gene transfer occurs at high rates in natural populations of bacteria and that genetic innovation is not restricted to extrachromosomal DNA. Additional conclusions, also contrary to old concepts, are that the bacterial genome can accommodate novel genetic material without challenging the nucleoid structure and that a larger genome is not necessarily a burden. Acquired DNA may immediately increase bacterial fitness or, more likely, remain as neutral genetic material until it is either turned useful or lost. Bacteria can thus play trial and error with horizontally acquired DNA. The ease with which bacteria incorporate foreign

DNA into the chromosome has truly surprised microbiologists. We are now inclined to view bacterial species as a group of isolates that share a more or less constant core or "backbone" genome but can show striking variations in their peripheral gene content.

Variation of Mutation Rates

Studies with clinical isolates have indicated that hypermutable bacterial lineages may adapt better to harsh environmental conditions. While it seems out of the question that such hypermutable lineages may enter an evolutionary dead end, their existence emphasizes the importance of mutation as an adaptive strategy. In fact, population geneticists have predicted that variation of mutation rates in response to environmental circumstances might have selective value. An increase of mutation rates during stressful conditions would increase genetic polymorphism, and diversification of populations might hedge their risk of disappearance. Under comfortable circumstances, however, elevated mutation rates would be unnecessary and probably detrimental.

An example of variation of mutation rates upon environmental influence is observed when *E. coli* is exposed to fluoroquinolones, a class of antibiotics that target DNA topoisomerases, thus blocking DNA replication. DNA replication arrest induces the SOS response, and DNA replication blockage is overcome by translesion synthesis DNA polymerases such as polymerase IV (Pol IV) (DinB) and Pol V (UmuDC). Because these alternative DNA polymerases, especially Pol V, are more error prone than DNA Pol III, the mutation rate increases. Of course, most of the mutations produced can be expected to be deleterious. However, the appearance of rare quinolone-resistant cells can permit survival of the population in the harsh circumstances created by the presence of the antibiotic. Increased mutation rates may produce additional mutations that can further facilitate survival.

SOS induction associated with antibiotic challenge is not the only environmentally controlled mechanism known to modulate mutation rates. Another example involves the RpoS-dependent general stress response, which is triggered by carbon source starvation and by a variety of stress signals. RpoS has been shown to downregulate MutHLS-mediated mismatch repair. An obvious consequence must be an increased frequency of base substitutions and other mutations eliminated by the MutHLS system. Furthermore, the RpoS sigma factor activates expression of the *dinB* gene, which encodes DNA Pol IV. Because Pol IV is error prone during replication of undamaged DNA, the mutation rate will increase.

Are the fluoroquinolone and RpoS examples exceptional or common? This question has been a matter of controversy for nearly two decades. One view maintains that variation of bacterial mutation rates may be a widespread strategy in response to stress. However, other authors have provided evidence that certain experiments supporting the occurrence of higher mutation rates under stress, including the famous 1988 article published in *Nature* by John Cairns and coworkers, contained subtle but fatal flaws. Furthermore, a major criticism based on theoretical calculations is that increased mutation rates might involve a mutational burden incompatible with adaptation.

Clues of Bacterial Evolutionary Success

Bacteria are equipped with analogue devices that permit efficient adaptation to changing conditions. Furthermore, bacterial populations often display bistable or multistable states, created either by built-in mechanisms or by random fluctuations. As with any other source of polymorphism, this heterogeneity can increase the chances that a bacterial population can cope with an environmental challenge. However, when we wonder at the adaptive capacity of a living bacterium, we are merely observing an image of a movie that started 3,500 billion years ago and moves very, very fast. Natural selection is as inexorable for bacteria as for any other living organism. However, the laws of bacterial variation are less stiff than in multicellular eukaryotes. Besides mutation and homologous recombination, acquisition and loss of DNA are additional modes of variation, whose importance in bacterial evolution has been understood very recently. Furthermore, the rates of bacterial mutation seem to vary under certain circumstances, thereby providing a further source of polymorphism. Whatever the origin of genetic change (mutation, rearrangement of preexisting material, or acquisition or loss of DNA), a major factor that speeds up bacterial evolution is that bacterial DNA is both somatic and germinal. As a consequence, favorable mutations are immediately passed to the offspring without the gamble of meiosis and gamete assortment.

FURTHER READING

Casadesus J, Low D. 2006. Epigenetic gene regulation in the bacterial world. *Microbiol Mol Biol Rev* **70:**830–856.

Dubnau D, Losick R. 2006. Bistability in bacteria. *Mol Microbiol* **61:**564–572.

Fitch WM. 1982. The challenges to Darwinism since the last centennial and the impact of molecular studies. *Evolution* **36:**1133–1143.

Fraser C, Alm EJ, Polz MF, Spratt BG, Hanage WP. 2009. The bacterial species challenge: making sense of genetic and ecological diversity. *Science* **323:**741–746.

Frost LS, Leplae R, Summers AO, Toussaint A. 2005. Mobile genetic elements: the agents of open source evolution. *Nat Rev Microbiol* **3:**722–732.

Galhardo RS, Hastings PJ, Rosenberg SM. 2007. Mutation as a stress response and the regulation of evolvability. *Crit Rev Biochem Mol Biol* **42:**399–435.

Gould SJ. 1994. The evolution of life on the earth. *Sci Am* **271:**84–91.

Haeusser DP, Levin PA. 2008. The great divide: coordinating cell cycle events during bacterial growth and division. *Curr Opin Microbiol* **11:**94–99.

Hendrickson H. 2009. Order and disorder during *Escherichia coli* divergence. *PLoS Genet* **5:**e1000335.

Janion C. 2008. Inducible SOS response system of DNA repair and mutagenesis in *Escherichia coli*. *Int J Biol Sci* **4:**338-344.

Kaern M, Elston TC, Blake WJ, Collins JJ. 2005. Stochasticity in gene expression: from theories to phenotypes. *Nat Rev Genet* **6:**451–464.

Kogoma T. 1997. Stable DNA replication: interplay between DNA replication, homologous recombination, and transcription. *Microbiol Mol Biol Rev* **61:**212–238.

Ochman H, Lawrence JG, Groisman EA. 2000. Lateral gene transfer and the nature of bacterial innovation. *Nature* **405:**299–304.

Prieto AI, Hernández SB, Cota I, Pucciarelli MG, Orlov Y, Ramos-Morales F, Garcia-del Portillo F, Casadesús J. 2009. Roles of the outer membrane protein AsmA of *Salmonella enterica* in the control of *marRAB* expression and invasion of epithelial cells. *J Bacteriol* **191:**3615–3622.

Prieto AI, Ramos-Morales F, Casadesus J. 2006. Repair of DNA damage induced by bile salts in *Salmonella enterica*. *Genetics* **174:**575–584.

Roth JR, Kugelberg E, Reams AB, Kofoid E, Andersson DI. 2006. Origin of mutations under selection: the adaptive mutation controversy. *Annu Rev Microbiol* **60:**477–501.

Saint-Ruf C, Matic I. 2006. Environmental tuning of mutation rates. *Environ Microbiol* **8:**193–199.
Smits WK, Kuipers OP, Veening JW. 2006. Phenotypic variation in bacteria: the role of feedback regulation. *Nat Rev Microbiol* **4:**259–271.
Waters LS, Storz G. 2009. Regulatory RNAs in bacteria. *Cell* **136:**615–628.
Wilson M. 2005. *Microbial Inhabitants of Humans.* Cambridge University Press, Cambridge, United Kingdom.
Zoetendal EG, Vaughan EE, de Vos WM. 2006. A microbial world within us. *Mol Microbiol* **59:**1639–1650.

Microbes and Evolution: The World That Darwin Never Saw
Edited by R. Kolter and S. Maloy
©2012 ASM Press, Washington, DC
doi:10.1128/9781555818470.ch15

15

The Impact of Differential Regulation on Bacterial Speciation

Eduardo A. Groisman

Members of a family can be quite different from one another, displaying distinct preferences for foods, exhibiting dissimilar tolerances to low or high temperatures, and showing diverse reaction times to a new situation or condition. Interestingly, this is true not only for humans and other animals but also for bacteria, the most abundant life form on earth. What, then, is the basis for the diverse behaviors that distinguish closely related organisms such that, for example, one can be extroverted and social, whereas a close relative prefers to be a loner, and what determines whether one organism establishes beneficial rather than parasitic interactions with another organism? Are the types of genetic differences that set apart closely related bacterial species similar to those that distinguish closely related animal species?

The family *Enterobacteriaceae*, whose members are often referred to as the enterics, comprises several bacterial species that are known because of the diseases they cause in humans and/or in

Eduardo A. Groisman received his Ph.D. from the University of Chicago and conducted postdoctoral research at the Scripps Research Institute and the University of California, San Diego. He was on the faculty at the Washington University School of Medicine for 20 years before moving to the Yale School of Medicine in 2010. He enjoys literature and music and spending time with his family.

animals or plants of economic importance. These include the gastroenteritis and typhoid fever bacterium *Salmonella enterica*, the bubonic plague agent *Yersinia pestis*, the bacillary dysentery-causing *Shigella flexneri*, and various *Erwinia* species, which are pathogens of potato and other crops. *Enterobacteriaceae* also comprises species that do not normally cause disease, many of which engage animal hosts in positive interactions. *Escherichia coli* is a normal member of the human intestinal flora. *Klebsiella pneumoniae* is often found in soil, where it can fix nitrogen, whereas *Buchnera aphidicola* establishes a partnership with plant pests known as aphids whereby *Buchnera* provides the aphid with certain essential nutrients and receives from it a different set of nutrients. What, then, distinguishes a pathogenic species from one that is normally innocuous?

While investigating the origins of pathogenicity, Howard Ochman, who is an evolutionary biologist presently at Yale Microbial Diversity Institute at Yale University, and I proposed four possible genetic scenarios to account for the differences that exist between the closely related bacterial species *S. enterica* and *E. coli*. First, *Salmonella* may harbor certain (virulence) genes that *E. coli* does not have. Second, *E. coli* may carry a virulence suppressor gene(s) that somehow interferes with the production, stability, and/or deployment of a virulence protein(s). Third, variations in the amino acid sequences of proteins that are shared between *Salmonella* and *E. coli* might contribute to a pathogenic personality in the former or prevent it from becoming a pathogen in the latter. And fourth, *Salmonella* and *E. coli* could differ in the ways in which they control when, where, and at what levels they manufacture the products they have in common. We now know that all four scenarios contribute to the phenotypic differences that set apart *Salmonella* from *E. coli*, not only enabling bacterial growth in locales within animal hosts but also allowing the utilization of particular nutrients and displaying resistance to certain antibiotics.

The first two scenarios are intuitive and straightforward: a pathogen is a pathogen because it has something that a related, nonpathogenic organism lacks, and vice versa. Species-specific regions of bacterial genomes often originate via a process known as horizontal (or lateral) gene transfer whereby one organism acquires a DNA segment from a different, often quite distant (evolutionarily speaking) organism that harbors the genes conferring a particular trait. For example, a bacterial species can become resistant to an antibiotic by obtaining the resistance genes from another species that is already resistant to the antibiotic. Horizontal gene transfer is rare in eukaryotic organisms, such as animals and plants, which develop new abilities typically by mutating their preexisting genes in ways that affect the properties of the encoded proteins or the circumstances under which particular proteins are produced.

The third and fourth scenarios can also give rise to significant behavioral differences, but their subtle genetic nature renders them more difficult to uncover. This is because small changes in the amino acid sequence of a protein can alter its function dramatically. In addition, transcription factors, which are proteins that govern the synthesis of the various proteins encoded in a genome, bind to short specific DNA sequences, and modifications of such sequences can have dramatic consequences by preventing a transcription factor from exerting its regulatory control. As transcription factors are produced and/or activated in response to particular signals, this can give rise to situations whereby a protein will be made in response to one signal in one species but not in another, closely related species because the latter lacks the DNA sequence that is recognized by the transcription factor. Likewise, small variations in a conserved regulatory protein, such as a transcription factor, can give rise to phenotypic differences among related organisms if the variant regulatory protein fails to recognize an interacting partner(s).

Let me illustrate how variation in gene regulatory strategies can result in distinct bacterial behaviors. Enteric bacteria differ in how they go about making the proteins that render them resistant to the antibiotic polymyxin B. The positively charged polymyxin B normally binds to the negatively charged surface of enteric bacteria. Not surprisingly, an enteric bacterium can become resistant to polymyxin B by decreasing the negative charge of its cell surface, thereby reducing polymyxin B binding. Changes in the bacterial cell surface charge are brought about by a group of proteins whose amino acid sequences are highly conserved. Enteric bacteria differ in the regulatory circuits that control where and to what levels the polymyxin B-resistant proteins are made. These comparative studies have revealed two distinct aspects that distinguish individual enteric species. First, organisms often vary in the environments in which the polymyxin B resistance proteins are produced, and this is often due to changes in the amino acid sequence of a regulatory protein. Second, two related organisms can synthesize the resistance proteins in response to the same signals but achieve dissimilar resistance levels as a consequence of the particular architecture that individual species utilize to promote expression of the resistance genes. In other circumstances, two regulatory architectures may differ in the time it takes to produce the polymyxin B resistance determinants once an organism experiences an inducing condition, or in how long the resistance proteins are made once the inducing condition is no longer present. Because polymyxin B is produced by a soil bacterium and because the cell surface modifications conferring polymyxin B resistance also enhance survival against toxic metals present in soil, it is likely that the distinct architectural designs that govern these modifications contribute to the ability of individual enteric species to survive in soil and, potentially, to occupy other niches.

How does the acquisition of DNA sequences by horizontal gene transfer affect a recipient bacterium? On the one hand, the acquired genes may enable the recipient bacterium to explore a new locale if they code for proteins that mediate entry and/or survival within a eukaryotic host, or that enable the recipient bacterium to utilize novel compounds as carbon and/or energy sources. It is often the case that the gene cluster mediating a new function, such as the capacity to invade animal cells, also includes genes coding for a regulatory protein(s) that controls the synthesis of the invasion proteins. Yet, even when a recipient bacterium acquires a regulatory gene(s) along with the invasion genes, it invariably relies on its ancestral transcription factors to govern the expression of the acquired genes. This is because the synthesis of the proteins encoded by horizontally acquired genes must be tightly regulated so that they are manufactured only when and where they are needed, and in a coordinated fashion with the proteins made by the recipient organism. In addition, to prevent the potential detrimental effects resulting from the inappropriate production of newly acquired proteins, bacteria often silence the expression of the acquired genes and rely on specific regulatory proteins to overcome silencing.

Besides the effects described above, a DNA segment incorporated as a result of horizontal gene transfer can affect the evolutionary trajectory of regulatory proteins encoded by genes that were present in the bacterium's genome before the incorporation of the new DNA took place. The DNA region controlling expression of a horizontally acquired gene frequently differs from that corresponding to an ancestral gene even when both genes are controlled by the same ancestral regulatory protein. This allows a given species to use the same transcription factor to differentially express proteins encoded by ancestral and horizontally acquired genes. When one compares the regulatory DNA regions controlled by a transcription factor present in two different species, one finds that the regulatory regions corresponding to ancestral genes tend to be similar, whereas those controlling horizontally acquired genes can be quite different. This can give rise to a situation where an ancestral transcription factor from one species can function in a different species to promote expression of ancestral, but not horizontally acquired, genes.

The types of genetic differences that distinguish closely related species can be quite different depending on whether one is considering bacterial versus eukaryotic species. Because different animal species have similar gene compositions, morphological and/or behavioral divergences are typically ascribed to changes in the regulatory regions of genes that are present in two related animals. Furthermore, animals are subjected to horizontal gene transfer only rarely, so their transcription factors are not normally involved in the control of novel sequences with distinct regulatory regions. By contrast, the high prevalence of gene acquisition from other sources has had an impact on bacterial genomes: related species not only display differences in the regulatory regions of conserved genes, like related eukaryotic organisms, but they also affect the transcription factors themselves. Of course, this is in addition to the new functions encoded by the acquired genes.

A significant proportion of eukaryotic genomes do not code for proteins and were deemed to be "junk" DNA. However, we now know that these genomic regions are functional, as they harbor the information to make a different nucleic acid, termed RNA, that can regulate the production of proteins in a variety of ways. It is becoming increasingly clear that many genetic diseases of humans are associated with mutations in the genes coding for regulatory RNAs. Bacterial genomes are compact (i.e., the proportion of the genome that does not code for proteins is very small). Yet, they also harbor regulatory RNAs that exert gene control by different mechanisms and affect their capacity to proliferate in various environments. It will be interesting to examine the evolution of RNA-mediated gene control and how selection operates on segments of bacterial genomes that appear to have multiple roles by coding for proteins and for regulatory RNAs.

Microbes and Evolution: The World That Darwin Never Saw
Edited by R. Kolter and S. Maloy

doi:10.1128/9781555818470.ch16

16

An Accidental Evolutionary Biologist
GASP, Long-Term Survival, and Evolution

Steven E. Finkel

The bacteria that come out of the test tube are different from the bacteria that went into the test tube. That is the essence of a phenomenon called the growth advantage in stationary phase (GASP), a phenotype resulting in an ability to observe evolution in a test tube in real time. As I start defining terms, I'll also explain how I came to study the evolution of bacteria...accidentally.

Within the laboratory, the bacterial life cycle consists of five phases: (i) lag phase, (ii) exponential or logarithmic phase, (iii) stationary phase, (iv) death phase, and (v) long-term stationary phase (LTSP). During the first phase, just after bacteria are inoculated into fresh culture medium, the cells have to "retool" their metabolism. They have most likely been either frozen or starved; either way, they are not equipped for rapid growth. Upon sensing the presence of nutrients, they begin to create the machinery that allows them to start reproducing rapidly. This period, one of abundant growth, is referred to as exponential phase (or

Steven Finkel earned his degrees at UC Berkeley and UCLA School of Medicine and was a postdoc at Harvard Medical School before joining the faculty at the University of Southern California in 2000. When not watching microbes evolve, he enjoys watching his twin daughters braid the flagella of giant plush microbes.

logarithmic phase) because the cells are growing so their numbers double at rapid rates, increasing by orders of magnitude in a relatively short period. For example, *Escherichia coli* (my primary microbe of study) can reproduce every 20 minutes during exponential phase. This means that if we introduce just a single cell into a test tube in the morning, we will have 10 billion cells by the afternoon. However, this kind of explosive growth cannot be sustained. Eventually nutrients are depleted, waste products accumulate, and the cells sense that they are at high density. At this point, cells enter the third phase, stationary phase, so named not because the cells stop swimming around but because the rapid increase in biomass that the culture has been experiencing stops. (A plot of growth as a function of time flattens out, so the graphed points appear "stationary.") Bacteria can remain at high density in stationary phase for quite a while; however, this phase also cannot last indefinitely. Eventually, and for reasons that are still not fully understood, cells begin to die, marking entry into the death phase. During death phase cells die rapidly; in fact, just as growth was exponential during exponential phase, loss of viability is also exponential during death phase. However, not all the cells die. For *E. coli* incubated under standard laboratory conditions, after about 99.9% of the cells die, the remaining cells can survive for extremely long periods (especially compared to the 20-minute generation time mentioned above). This signals entry into the fifth phase, which we call LTSP. One textbook from 1935 more poetically refers to it as the "phase of prolonged decrease." However, unlike the first (stationary) phase, where there is little cell growth, LTSP is a period of dynamic change.

And this brings me to how I came to study bacterial evolution. My research interest as a new postdoctoral fellow was to study how the bacterial chromosome is packaged during the stationary phase of the bacterial life cycle. It was clear that during the transition out of exponential phase many important changes were occurring in the cell's DNA and that these changes were essential for the survival of the organism. I was working in the laboratory of Roberto Kolter at Harvard Medical School, and his lab was pioneering the study of the control of gene expression during stationary phase. At that time, there was increasing appreciation for the fact that in the real world, microbes probably did not experience the kinds of lifestyles we gave them in the laboratory. In the lab, cells are generally nurtured under conditions optimized for rapid growth; microbiologists (myself included) love to study exponential-phase cells...it's so easy! However, in the real world, life for most microbes is far more difficult. Stationary phase was our way of beginning to emulate these more natural conditions.

Although it was well established by Roberto's group that interesting things happened to bacteria that were incubated for 10 days (which at the time was an almost unheard-of length of incubation time when studying *E. coli*), it turns out that bacteria could be maintained even longer, without the addition of any new nutrients. I came to learn of this almost by accident (accidents are a recurring theme in this essay). A graduate student in the lab, Sara Lazar (now a neurobiologist), mentioned in passing that bacteria could survive a "long time" in the lab, much longer than 10 days. I asked, "How long?" and Sara replied, "A really long time." I asked again, "How long is that?" She replied, "A really, really long time." Sometimes this kind of interaction is all the impetus a scientist needs to start an experiment that will forever change the direction of his or her research.

Later that day, I started six cultures designed to test the idea of how long a "really, really long time" might be. At first I checked on the viability of the cells every day, then every 10 days, then once a month. Every time I checked, not only were the cells still alive, but they were present at the same cell density: approximately 5 million cells per milliliter of culture. These cultures survived for more than 5 years without my adding any nutrients. (In fact, the cultures would probably still be going today, almost 15 years later, had the experiment not been terminated when it was time for me to move to my own laboratory as an assistant professor at the University of Southern California.) So, what were these surviving bacteria eating? For each cell that survived, about 1,000 cells must have died; the bacteria could recycle the nutrients in the test tube and would not die out completely. However, very soon after the experiment started I began to notice that some of the bacteria living in the culture were different from the original parents in the inoculum. *E. coli* organisms have a distinctive size, shape, color, and even smell when they are spread onto petri dishes containing nutrient agar (think of it as Jell-O mixed with Vegemite) and allowed to form colonies. After just 60 days of growth in the test tubes, the bacteria I sampled began to show altered appearances. We call these variations different colony "morphotypes." The morphotypes varied in size (some larger and many smaller), color (whitish or yellowish instead of the usual beige color), and the production of extracellular substances. Some of the colonies were gooey (formally called "mucoid"), and others were a bit wrinkly (called rugose). My first thought was that these foreign-looking cells were contaminants, accidental invaders from either my own hands or the ventilation system of our venerable building on the med school quad. However, several genetic tests (later confirmed by more exacting molecular biological techniques) demonstrated unequivocally that these cells not only were *E. coli* but also, remarkably, were *E. coli* descended from the same parent.

The idea that mutations could occur in test tube cultures and undergo natural selection had been known for some time. Research spearheaded in the Kolter lab had shown that after only 10 days, mutants could be isolated that expressed a significant competitive advantage over the parental strain, expressing what came to be called the GASP phenotype; GASP described something quite astounding. If one simply removes a small number of cells from a 10-day-old culture and introduces them as a minority into a 1-day-old culture, the cells from the older culture begin to take over! After a few days, we see an increase in the number of cells from the aged minority population as they steadily become the majority and drive the younger population to extinction. Cells from 10-day-old cultures will always take over cells from 1-day-old cultures; cells from 20-day-old cultures always overtake cells from 1- or 10-day-old cultures, etc.... So, what is happening? Where do the mutants come from, and why are they more fit than their parents and able to take over the culture?

In order to appreciate where the mutants come from, we need to understand the origins of genetic mutation...another series of accidents. To reproduce, all organisms need to copy their DNA and pass on their genes to the next generation. Bacteria like *E. coli* have just one chromosome of about 4.6 million base pairs, with about

4,300 genes. Each generation, the cell makes a duplicate copy of its DNA, grows to about twice its normal size, separates the two copies, and splits down the middle (this is formally called cell division by binary fission). The two new daughter cells are almost always perfect copies of the original mother cell...almost always. The process of copying DNA, while remarkably accurate, is not perfect. Mistakes in the copying process occur approximately once for each 10 billion bases of DNA replicated. However, the genome of a typical bacterium is only millions of bases in length. The practical result of this high degree of reproductive accuracy is that in *E. coli*, for every 10,000 new cells created, 9,999 are perfect copies of the original great-great-great-great (great × 12) grandmother and 1 cell will be a mutant that differs at a single location. One base pair in the entire genome of that 10,000th cell will be different than in all her sisters and cousins. However, that one change in the genetic code can spell the difference between being more fit than everyone else or going extinct. This is what GASP is all about.

We often view mutations as bad things. Certainly in science fiction movies the mutants are rarely the "good guys" (X-Men being a notable exception). In Darwinian terms, mutations can be either beneficial, neutral, or deleterious. Deleterious mutations will reduce an organism's fitness, and those microbes will disappear from a population or laboratory culture; neutral mutations may or may not propagate through a population, but do so for reasons other than their contribution to fitness; but beneficial mutations can significantly increase a cell's ability to grow and reproduce. GASP mutations fall into this last category. They confer an ability on the cells that their sisters do not possess, allowing them to increase their relative rate of reproduction, which increases their frequency in the population, such that what was once a single cell minority can become a million-fold majority. And, this can happen very quickly!

In a typical GASP experiment, a culture is aged for 10 days. During this time millions of random mutations arise. Among these millions are hundreds, possibly thousands, of mutations that are actually beneficial to the cell. However, it is very important to remember that the cell doesn't "know" that these mutations might be beneficial and did nothing intentional to help them to appear; it is all due to an accident—the reproductive machinery made a mistake and there happened to be an advantageous outcome. The same process that gives rise to beneficial GASP mutants could just as easily lead to the death of the cell. When some of these aged cells are sampled and transferred into a test tube containing only unaged, parental cells, the few GASP mutants that got transferred now have the opportunity to express their relative increase in fitness. To give an example, frequently GASP mutants can eat things that the parental strain doesn't eat, or eats them more efficiently. To date we have seen GASP mutants consume amino acids and proteins better than their unevolved siblings; some mutants might even eat DNA! My research now focuses on understanding the mechanisms that lead to the generation of this genetic diversity and the myriad ways cells can increase their relative fitness.

One of the challenging concepts to grasp when thinking about GASP, or any evolutionary process, is the fact that the kinds of events we are talking about (for example, the single DNA mutation event that leads to a beneficial trait) are very rare, happening at frequencies of one in a billion. However, an important fact to keep in mind is that in a typical bacterial culture there can be billions of cells and quadrillions of base pairs of DNA. While an accident like a beneficial mutation is rare, there are enough individuals present in the population to regularly experience such accidents. As humans we almost never consciously encounter billions of anything, so we don't typically process incredibly large numbers. We have to "retool" our thinking to appreciate this. The closest analogy on the human scale I can come up with is something like this: we often say that if there's no chance of something happening to someone, the chance is "one in a million." But, that means it's happening to eight people in New York City right now! When we consider that the population of one test tube is the equivalent of 1,000 New York Cities, the potential amount of diversity is staggering.

So, that's how I became an unofficial evolutionary biologist. One simple question compelled me to watch populations of bacteria in test tubes do things that they were not "supposed" to be able to do in the laboratory. We now know that these test tube universes are far more diverse than we had previously imagined. The late evolutionary biologist and author Stephen Jay Gould used to talk about "replaying the tape" as a way to imagine how evolution can proceed. In his analogy, the history of the Earth is being recorded (most likely on something that looked like a reel-to-reel tape recorder), from before life appeared to the present day. Steve argued that if you rewound the tape to a point just before life appeared and then started the tape forward again, life would again evolve. However, it might not be the same kinds of life as we know today. Due to chance and contingency (accidents!), a mutation that had occurred in one place at one time could now occur elsewhere. So, life would evolve along a new trajectory but still follow the rules of Darwinian natural selection.

Obviously, we can't replay the tape. However, the GASP phenomenon allows us to do something almost as good: we can set up many, many tape players running in parallel. We can create true microcosms, tiny universes, where in the beginning all conditions are identical. Experiments like these allow us to explore the mechanisms that lead to the generation of genetic diversity, the raw material of natural selection, and how evolutionary processes, as envisioned by Darwin, continue to shape the biosphere.

Microbes and Evolution: The World That Darwin Never Saw
Edited by R. Kolter and S. Maloy

doi:10.1128/9781555818470.ch17

17

How Bacteria Revealed Darwin's Mistake (and Got Me To Read *On the Origin of Species*)

John R. Roth

Classic science papers and books are often cited but seldom read. I confess, I worked on bacterial genetics for 30 years before I actually sat down with Darwin's magnum opus *On the Origin of Species*. A question about the origin of mutations finally drove me to the original tome. We have been engaged in a debate on the origins of mutation, and our conclusions contradicted something attributed to Darwin. We really wanted him on our side—Darwin knew that his theory of natural selection requires a steady injection of new changes in the information set (Darwin called it natural variation). Without this influx, selection will remove the bad versions of each gene, leave every organism with the best available version, and then come to a halt—game over, no more change, you are dead in the water. It was understandably difficult for Darwin to imagine the source or nature of this variation, because he wrote *Origin* without knowing anything about inheritance—how genetic infor-

John Roth was a Ph.D. student with Phil Hartman and a postdoctoral fellow with Bruce Ames. He has taught at UC Berkeley, the University of Utah, and UC Davis. His lab has worked on use of transposons in genetics, on metabolism (histidine, NAD, B_{12}), gene regulation, and chromosome rearrangement, and most recently on the origin of mutations under selection.

mation is stored, copied, and transmitted. Above all, he did not have any sharp idea about how errors arise and how a change in information affects form and behavior (phenotype). He knew that all these things were happening; he just did not know how. The source of mutations was a particularly hard problem. Three chapters of *Origin* are dedicated to the question of variability.

The origin of mutations is still a problem—150 years later—despite all we know about DNA structure, base sequence information, genetic coding, and the molecular nature of mutations. The question remains as "pesky" as it was for Darwin, because it is difficult (even now) to measure the rate at which mutations arise, especially those that affect growth ability (the only interesting kind). There are many mutation types, each arising at a characteristic frequency and affecting phenotype in a particular way. When a mutation arises in a natural population, the mutant organism is immediately subject to natural selection. A beneficial mutation enhances reproductive success and increases its frequency in the population. A bad mutation impairs reproduction and reduces its frequency even to the point of elimination. The effects of selection on mutant frequency obscure the rate at which mutations (good or bad) arise. An increase in mutant frequency could reflect an increased mutation rate or an altered growth rate of the mutant organism. If you cannot measure mutation rate, how can you identify the cause of the mutation? Are mutations made at a constant steady rate? Are they made at a higher rate when outside conditions signal the need for a change? Can cells preferentially make useful mutations and avoid the bad ones?

Darwin knew that the evolutionary changes he observed in nature occurred most often by a series of small changes. He also knew that mutants with large phenotype changes appeared in domesticated plants and animals (freaks or sports) but were seldom seen in natural populations. It must have looked to him as if huge, bad things happen while we are watching (mutations with large nasty consequences) and good things happen under cover of darkness (small-effect beneficial mutations in natural populations)—a very peculiar situation. He suggested that stress or "need" might lead to an influx of variation (mutations: some good and some bad) and that selection operated on this range of types to favor changes that improved reproduction (the essence of natural selection). He suggested that organisms cultivated by humans are subject to the stress of being taking from their natural environment and that this might explain the appearance of more (or bigger-effect) mutants. (You might, of course, also argue the converse—that stress is reduced when domesticated organisms are protected from the rigors of "life in the wild.") Without a theory of inheritance, Darwin had no way to know the formation rates of mutations with large or small effects on phenotype or the relative frequency of mutations that improve or impair reproductive success. Without a way to measure rates, it is hard to know whether those rates are regulated in response to stress, but the question is fundamental.

Darwin really could have used bacteria, organisms for which you can really measure mutation rates—if you are careful. Astronomical populations of bacteria can be grown with and without stress and their mutation types can be characterized. In fact (100 years after Darwin), bacteria helped Lederberg, Luria, and Delbrück show that new mutations can form without any required "stress." Mutants arise continuously in perfectly happy bacterial populations. They showed that you can measure mutation rates if you eliminate the effects of selection. Long before I got interested in science, most people had already accepted their classic evidence that "mutations are random errors in replication or repair of DNA." Selection can detect mutants that arise in the population before any stress is imposed. It seemed as if Darwin had made a mistake—mutations are not caused by stress. As a graduate student, I assumed that this question was answered. But now, 30 years later, the same question has resurfaced and is driving us crazy.

My first hint that Darwin's view of mutation might be resurrected came 20 years later when I visited John Cairns at Harvard Medical School. With great excitement he showed me his data (obtained using bacteria) suggesting that Lederberg, Luria, and Delbrück had missed something—Darwin might have been closer to the mark than people realized. The problem with the classic experiments, Cairns said, was that Lederberg, Luria, and Delbrück had all used lethal selections that detected mutants as colonies on petri plates and showed that these colonies were initiated by mutant cells that arose well before exposure to selection. When nonmutant parental cells hit the selective plate, they died before they could respond, making it impossible to detect mutations induced by selective conditions. Maybe, Cairns suggested, the early experiments detected only mutations that arise at random before selection and missed another sort of mutation that is generated in response to stress; the latter type escaped detection because it was killed by lethal selective conditions. Cairns further suggested that cells might direct their "homemade" stress-induced mutations preferentially to sites whose alteration could improve growth. He supported these suggestions using a nonlethal selection regimen that allowed stressed cells to respond to selection. The key to evaluating his claims lies in the details of his selection conditions.

Cairns' experiment is basically simple. Mutant bacterial cells unable to use lactose as a carbon source (Lac) are spread on lactose medium. The bulk population cannot grow, but a lot of mutant (Lac^+) colonies appear over a period of 6 days. Each colony is said to be due to a large-effect mutation (Lac^- to Lac^{+++}) occurring in the nongrowing population. The accumulation of normally rare Lac^+ mutants over time is attributed to mutagenesis of the nongrowing population during prolonged exposure to selective stress. Cairns' experiment seems to vindicate Darwin—mutations seem to be induced by stress. Had the classic experiments missed something, or was there something wrong with Cairns' system?

Having a healthy respect for the classics, we assumed that something must have been wrong with Cairns' experiment. How could cells "know" what mutations they needed and make them to order? If cells could actually direct mutations in a purposeful (nonrandom) way, evolution could occur without natural selection. Mathematical modeling suggested that random mutagenesis is not a prudent means of self-improvement—the target for bad mutations is just too big. A random mutation is at least 10,000 times more likely to do harm than good. (It is like fixing your watch with a hammer.) Would a cell put its genetic heritage in jeopardy, just because it has a "bad hair day"? How can a cell be sure that the problem (e.g., a lack of food) can even be solved by mutation? Slowly, over the last few years, a picture has emerged that makes sense of Cairns' experiment in terms of natural selection (not mutagenesis). That is, we think that Darwin was wrong about stress-induced mutation, but his main idea—natural selection—can produce effects that look both mutagenic and omniscient. Before evaluating the Cairns experiment, let's take a look at the "small-effect" mutations that Darwin focused on; these are often overlooked in laboratory genetics and are central to this discussion.

The Importance of Small-Effect Mutations

Two points have emerged from molecular studies of mutation. Small-effect mutations are extremely common, and they can contribute serially to very fast strain improvement.

Small-effect mutations arise at a high rate. There is an inverse relationship between the rate at which a mutation type arises and the magnitude of its phenotypic effect. This is clear from work on the nature of various mutation types, the process of protein synthesis, and the structure of the genetic code. Only recently has it been shown that our genetic code was shaped by strong selection to minimize the phenotypic consequences of the commonest mutations. The commonest point mutations are transitions (A/T to G/C), of which about one-third cause no change in the encoded protein (synonymous). Many more of these mutations substitute a very similar amino acid for the original one. In contrast, mutations that cause translational reading frame changes (+1, −1) have serious consequences for gene expression (typically a 1,000-fold reduction in expression) and are much rarer than base substitutions. It seems that the mechanisms for DNA replication and repair have evolved to be most effective at preventing mutation types with large phenotypic effects. The most frequent mutations of all are changes in gene copy number (duplications and amplifications). Duplications in any particular gene arise at a rate of 10^{-5} per cell per division in *Salmonella* (and probably many other bacteria). Further changes in gene copy number (up or down) occur at a rate of about 10^{-2} per cell per division. Perhaps more than 10% of new chromosome copies carry duplications somewhere. The commonest mutations (small-effect amplification steps) occur at a rate nearly 1 million times that of the rarest (large-effect +1 frameshifts) (Fig. 1).

Serial improvement occurs by frequent small-effect mutations. The most frequent (small-effect) mutation types (e.g., duplications and amplifications) are the first to respond when a bacterial population is exposed to selection for an increase in gene activity. These mutations are likely to already be present in even small populations. Selection causes an exponential increase in the frequency of these mutants until the growing subpopulation can realize another improving mutation. A series of such events can occur in rapid succession, leading to quick improvement. When selection is imposed on solid selective medium (a petri dish of agar medium), then the whole improving lineage is localized to one spot on the plate and a common small-effect cell type can improve until a large healthy colony appears, most of whose component cells are substantially improved (Fig. 2).

Before You Can Estimate the Mutation Rate, You Must Eliminate Natural Selection

Like the classic experiments, Cairns' system is designed to estimate mutation rates—only now the mutations are being selected in a nongrowing stressed population rather than a culture growing without selection. To estimate the mutation rate, one must be sure

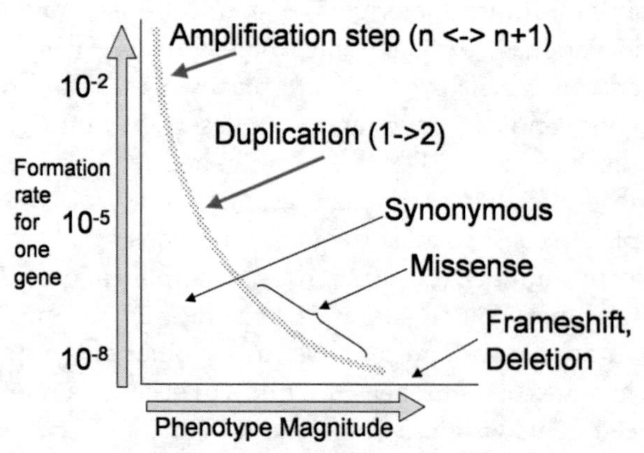

Figure 1 Mutation types vary in their formation rate over a millionfold range. In general, the most frequent mutation types are those with the smallest effects on phenotype. This is true for both increases and decreases in gene activity. doi:10.1128/9781555818470.ch17f1

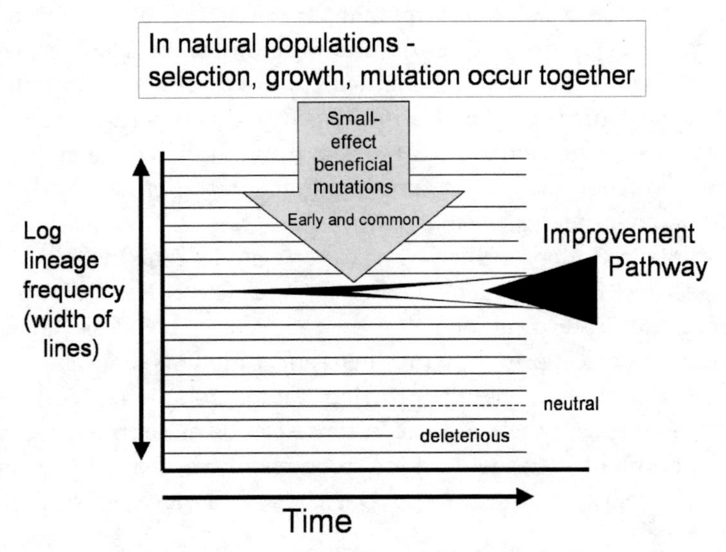

Figure 2 Selection causes an exponential increase in mutant frequency and is most likely to act on the most common mutations, which often have the smallest phenotypic change. Each mutation leads to expansion of a subclone in which subsequent common mutations can further enhance growth ability. doi:10.1128/9781555818470.ch17f2

that natural selection (differential growth) is not influencing mutant frequency. Luria and Delbrück used a very strong (lethal) selection to measure the rate of mutation during the preceding nonselective growth period. All of laboratory bacterial genetics uses the same principle—strong (but not always lethal) selection. The trick is to block all parent cell growth (no new mutations) and prevent growth of those annoying frequent small-effect mutations (which can improve rapidly under selection). This trick allows you to eliminate effects of natural selection and detect rare big-effect mutations whose frequency is dictated entirely by mutational events.

Why are Mutants Frequent in the Cairns Experiment?

The Cairns experiment looks like a lab experiment with one fatal flaw. True, selection for growth on lactose does prevent growth of the parent *lac* cells. However, selection is too weak to eliminate those sneaky small-effect mutants. Even a twofold increase in enzyme level (from a *lac* duplication) can allow slow growth. Nearly 1% of the cells in the parent culture carry a *lac* duplication. Over the course of 6 days—a prolonged selection period—many of these struggling clones manage to go through the whole multistep process—one small-effect amplification step after another—and ultimately generate a visible colony.

Interpretations of the Cairns experiment assume that only rare large-effect mutants (Lac^{---} to Lac^{+++}) are detected and that these are made more common by stress-induced mutagenesis. In fact, selection detects primarily the small-effect mutants that manage to become fully Lac^{+} while growing under selection for 6 long days—an eternity when division times can be as short as an hour. Formation of a visible colony requires more than 20 generations—plenty of time to use Darwin's main idea to solve the

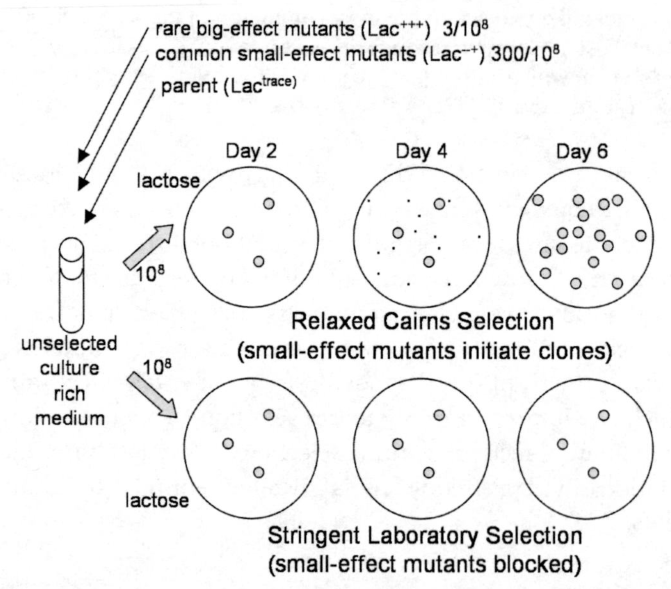

Figure 3 The Cairns experiment allows slow growth of very common small-effect mutants (with a *lac* duplication). These mutants initiate clones that adapt by a multistep pathway during the prolonged (6-day) selection period. Many of these clones succeed in generating full-sized Lac^{+} colonies, which first appear at various times over the selection period. doi:10.1128/9781555818470.ch17f3

problem. The colonies look like big-effect mutants induced by stress; they are actually very common small-effect mutations that improve as the colony forms. (This is diagrammed in Fig. 3.)

Take-Home Points

So my reading of *Origin* revealed both good news and bad. Darwin's idea of stress-induced mutation may be wrong, but natural selection can take on breathtaking power when common small-effect mutations are allowed to contribute. These effects are revealed by bacterial populations. Under selection, the high speed of genetic adaptation is easily mistaken for an increase in mutation rate—maybe even Darwin underestimated the power of selection. Readers should be aware that despite my confidence that the above interpretation is right, the idea of stress-induced mutagenesis is still strongly defended. Science is funny that way.

FURTHER READING

Cairns J, Foster PL. 1991. Adaptive reversion of a frameshift mutation in *Escherichia coli*. *Genetics* **128:**695–701.

Cairns J, Overbaugh J, Miller S. 1988. The origin of mutants. *Nature* **335:**142–145.

Darwin C. 1859. *On the Origin of Species.* John Murray, London, United Kingdom.

Foster PL. 2007. Stress-induced mutagenesis in bacteria. *Crit Rev Biochem Mol Biol* **42:**373–397.

Galhardo RS, Hastings PJ, Rosenberg SM. 2007. Mutation as a stress response and the regulation of evolvability. *Crit Rev Biochem Mol Biol* **42:**399–435.

Lederberg J, Lederberg EM. 1952. Replica plating and indirect selection of bacterial mutants. *J Bacteriol* **63:**399–406.

Luria SE, Delbrück M. 1943. Mutations of bacteria from virus sensitivity to virus resistance. *Genetics* **28:**491–511.

Roth JR, Kofoid E, Roth FP, Berg OG, Seger J, Andersson DI. 2003. Regulating general mutation rates: examination of the hypermutable state model for Cairnsian adaptive mutation. *Genetics* **163:**1483–1496.

Roth JR, Kugelberg E, Reams AB, Kofoid E, Andersson DI. 2006. Origin of mutations under selection: the adaptive mutation controversy. *Annu Rev Microbiol* **60:**477–501.

Microbes and Evolution: The World That Darwin Never Saw
Edited by R. Kolter and S. Maloy

doi:10.1128/9781555818470.ch18

18

The Role of Conjugation in the Evolution of Bacteria

Fernando de la Cruz

Conjugation is a sophisticated mechanism of DNA transfer between bacteria. Together with transformation and transduction, it substitutes for sex in prokaryotes. Usually, conjugation machineries are built on autonomously replicating units, independent of the chromosome, called plasmids. Apparently, the gene repertoire of plasmids is different from those of main chromosomes, so the term "mobilome" was coined to stress this fact. The mobilome is the genetic repertoire available to a given bacterium by lateral gene transfer, which is shared by a bacterial population. Thus, a bacterium like *Escherichia coli* contains 5 Mb of chromosomal DNA plus maybe 20 Mb or more in easily accessible information in the mobilome. As soon as there is any selective pressure, the relevant mobilome units are activated and they multiply in the population. This is the way multiple antibiotic resistance determinants have spread among human pathogenic bacteria in the last 50 years. The carriers of the resistances were transmissible plasmids.

There are many transmissible plasmids, but despite their diversity, they occur in just two basic formats. Conjugative plas-

Fernando de la Cruz has been Professor of Genetics in the University of Cantabria (Spain) since 1984. He is an expert in mobile genetic elements, with a focus in plasmid biology and the mechanism of bacterial conjugation. He has recently become a fan of synthetic biology.

mids contain a complete set of machinery for conjugative gene transfer (a minimum of about 15 genes or 20 kb of DNA). Mobilizable plasmids, on the other hand, contain just the essential requirements for conjugation, that is, the origin of transfer (called *oriT*)—a small DNA segment of 60 to 300 bp that is the site where conjugation starts and ends—and the gene for the relaxase (the protein that recognizes and interacts with the *oriT*). As a consequence of this simple fact, mobilizable plasmids are small (<15 kb) while conjugative plasmids are large (>30 kb). Plasmid size brings in other important consequences. If a plasmid is small, it can be in high copy number without representing an unacceptable burden for the cell. Thus, small mobilizable plasmids are usually high copy number, and they can be stably inherited by relying on random partition. To ensure their stability, they need a multimer resolution system to avoid the so-called dimer catastrophe (dimers and higher multimers are preferred to monomers as the selected unit for plasmid replication; thus, monomers are progressively converted to dimers and higher multimers in the absence of a multimer resolution system). In contrast to small mobilizable plasmids, large conjugative plasmids cannot be in high copy number, since that will bring their DNA mass close to that of the chromosome. Therefore, large conjugative plasmids must have a low copy number. This imposes additional genetic requirements, such as mechanisms for stability. The most widely spread stability determinants are the addiction modules (also called toxin-antitoxin systems) and/or the partition systems. In addition, plasmids, either in the mobilizable or in the conjugative version, contain a small number of adaptive genes that allow them to thrive under specific selective conditions (environments with antibiotics or strange toxic organic compounds, colonizing a eukaryotic host, etc.).

There are apparently different mechanisms to achieve conjugation if one judges by the lack of sequence similarity among conjugative plasmids. However, from the mechanistic point of view, most of them are related. Research has shown that there is a predominant mechanism of gene transfer by conjugation. This was a surprise, but a nice surprise, since all conjugative systems therefore use variants of a common conjugative processing mechanism. Let me describe it briefly.

The process of conjugation can be conveniently divided into three basic steps: initiation (or DNA processing in the donor), DNA transport, and termination (or DNA processing in the recipient). The initiation step involves an interaction between the conjugative *oriT* of the plasmid in the donor bacterium and the relaxase. The DNA is cleaved at a specific site within *oriT*, with the subsequent formation of a nucleoprotein complex (the T-complex). The T-complex contains a covalent bond formed between the 5′-phosphate of the cleaved DNA strand and a specific tyrosine residue in the relaxase. In the second step, the relaxase is taken up as a secretion substrate by a type IV protein secretion system, which forms the conjugation channel. The protein is transported out of the donor and into the recipient cell. Since the DNA is covalently bound to the relaxase, it is threaded into the channel. The transport process of the DNA is aided now by a motor protein, called coupling protein. Finally, the third step occurs in the recipient cell. When a complete copy of the donor DNA has been transferred, the relaxase can effect a second cleavage reaction, accompanied by a transesterification, which ligates the transported DNA molecule into a circular DNA and thus reconstitutes a complete copy of the plasmid in the recipient cell. This model, basically, proposes that the relaxase is shot to the recipient cell and then the DNA is pumped in. Thus, we called it the "shoot and pump" conjugation model.

The plasmid R388 is an ideal model to study conjugation because of its simplicity. Since our initial efforts in 1990, we have gained a lot of understanding both in the biochemical and in the structural sides of conjugation. R388 is among the best-known plasmids, as judged by the accumulated genetic, structural, and biochemical data.

My scientific life has revolved around bacterial mobile genetic elements (plasmids and transposons). In my group in Santander, we chose to study the mechanism of conjugation using plasmid R388 as a model. In the 1980s I was working on Tn*21* transposition in collaboration with John Grinsted of the University of Bristol. We were using plasmid R388 as a transposon recipient. R388 was convenient since it is a relatively small conjugative plasmid and, at the time, its small size helped restriction analysis for transposon mapping. I realized that R388 must have a small region involved in conjugative transfer since, in spite of being just 34 kb long, many of the recombinant plasmids containing transposon insertions were still conjugation proficient. Thus, my group in Santander started to analyze the mechanism of conjugation of plasmid R388. Soon we were excited to find that the conjugation genes of R388 were similar to those determining T-transport from *Agrobacterium tumefaciens* to plant cells. The fact that conjugation and DNA transport to plants use similar transport mechanisms had important implications, both on theoretical and on practical grounds. Theoretically, it expanded the importance of conjugation as a mechanism of DNA transport with a very broad range of potential recipient cells. In a more applied work, it gave conjugation the potential for reaching eukaryotic cells as targets. It was later shown that conjugation can occur with yeast and even animal cells. Interkingdom conjugation is now an active field of research.

With the generalization of DNA sequencing, it was found that many chromosomes contained integrated plasmid or plasmid-like sequences. The same occurred with phages, the other "classical" mobile genetic elements. Apparently, plasmids and phages landed in the chromosome, integrating novel genetic information and then undergoing a slow process of degeneration, to end up in progressively harder-to-recognize genomic islands of horizontal gene transfer origin. It was becoming more and more clear that important events in microbial speciation occurred by horizontal acquisition of new genetic information. By extrapolating this idea, in a sabbatical year in Vancouver (Canada) with Julian Davies, we arrived at the idea of horizontal gene transfer as the mechanism of formation of new bacterial species. In the scenario we envisaged, plasmids are the cavalry in the army of bacterial evolution, the first to arrive to the battlefield. Thanks to the efficiency and promiscuity of conjugation, plasmids can land anywhere from anywhere in very little time, thus bringing in the genetic repertoires that are required to face the novel selective pressures. However, after the first moment (hundreds or thousands of years, probably), their genetic instability and selfish tendencies make them far from the ideal loci for the now-successful adaptive genes. Apparently, these genes are stabilized by inclusion in genomic islands, silent remnants of the ancient mobile genetic elements. Nevertheless, not all plasmids disappear in the long run. Some become more and more adapted to a host, like the megaplasmids of *Rhizobium* and *Agrobacterium,* which are essential constituents of the genetic makeup of their hosts, to even chromosome 2 of *Vibrio cholerae,* an ancient plasmid now converted to a chromosome. Thus, conjugation plays a central role in reshaping chromosomes as a response to a selective pressure.

With time, work in my lab has progressively turned to study the inhibition of conjugation as a means to curb the spread of antibiotic resistance. The rationale behind this idea is that by inhibiting conjugation we are using a two-pronged weapon to combat bacterial disease: on one side we avoid the spread of antibiotic resistance, and on the other, we attack directly some mechanisms of virulence that directly use type IV secretion systems. In this vein, we have identified compounds that can act as conjugation inhibitors (we call them COINs, implying that they could be commercially attractive). In a first survey, financed by the European Union V Framework Programme, we screened several compound libraries by using high-throughput screening methods. We discovered that a number of polyunsaturated fatty acids were good inhibitors of plasmid R388 conjugation. However, none of the inhibitors found was of broad range, that is, was active against a series of plasmids representing the natural diversity. Therefore, our initial search was rather unrewarding. In order to find useful targets for the potential COINs, we turned to more rational approaches. We wanted to determine what *E. coli* genes are involved in good conjugation recipient ability. The surprise was that in standard *E. coli* laboratory strains, there are apparently no genes that can be mutated to impede conjugation. Conjugation (unlike bacteriophage infection) is therefore a process that cannot be easily inhibited from the recipient side. Why is this? We believe that conjugative plasmids, at least those with a broad range of prospective hosts, have carefully selected a mechanism of conjugation that forcefully infects new hosts and gives them little chance to avoid infection. By avoiding immunity, they are effectively shared by the most often complex bacterial ecosystem populating a given niche, and expand the reach of the mobilome.

In summary, conjugation is an essential feature in the physiology of plasmids. Being energetically costly, it engraves its constraints in the plasmid genetic makeup. Conjugation plays an important role in the evolution of bacterial genomes. This is especially true for short-term bacterial adaptation to new selective pressures. Human pathogenic bacteria have acquired many of their virulent traits, as well as antibiotic resistance, by conjugation. If we learn how to control conjugation, we might have an additional potent weapon to combat human bacterial disease.

Microbes and Evolution: The World That Darwin Never Saw
Edited by R. Kolter and S. Maloy

doi:10.1128/9781555818470.ch19

19

Do Bacteria Have Sex?

Rosemary J. Redfield

Do bacteria have sex? Most people probably think I'm nuts, but getting paid to answer this question makes me feel very lucky. Well, I'm also paid to teach biology and supervise graduate students and sit on committees, but this is the best part of my job. Why would anyone care whether bacteria have sex? Because the answer addresses our own most fundamental attributes—our sexuality and our reproduction.

The word *sex* has lots of different meanings. Do bacteria have a tiny form of sexual intercourse? No.... Do they come in males and females, with tiny private parts? Um, no.... My question asks about a more broadly applicable meaning of the word sex, *sexual reproduction*. Biologists are profoundly embarrassed by the existence of sexual reproduction, not because it's rude or kinky but because it doesn't make evolutionary sense. People who aren't biologists think it's perfectly sensible to make children by mixing sets of genes from two people. After all, how else would we do it? But to a biologist this seems very peculiar, because we know that the essence of reproducing is getting our own versions of genes into the next generation. That's how natural selection works—whatever versions of genes reproduce more efficiently

Rosie Redfield followed her Stanford Ph.D. with postdoctoral work at Harvard and Johns Hopkins, and has been at the University of British Columbia since 1990. She discusses her ongoing research in her RRResearch blog, which recently became famous for critiquing NASA's arsenic-DNA claims. Her latest non-science goal is to deadlift her own weight.

become more common over the generations—so replacing half of our own genes with half of someone else's genes should be a dead loss as an evolutionary strategy. Yet almost all plants and animals reproduce this way, and many of our single-celled relatives too, so biologists suspect that it must have benefits we've overlooked. We know that "asexual" reproduction is possible because new strawberry plants can grow from runners and whole starfish from broken-off arms.

The sexual way of reproducing has two steps the asexual way lacks: discarding one set of genes and mixing the remaining set with a set from someone else. We each have two versions of each gene, one from Mom and one from Dad, giving us one complete gene set from each parent. Our bodies remix them like two decks of cards before putting one mixed set into each egg or sperm, so that we give each of our kids a different mixed set of the two sets we inherited. Biologists know a lot about how the mixing and discarding happens—the fusing of egg and sperm, the molecular dance of the chromosomes when sex cells divide—but we haven't been able to figure out why. We do know that the explanation must lie with natural selection—the genes that cause us to reproduce this way must have evolved because they and our other genes were more successful when sexual reproduction mixed them with other versions. But that's really only restating the question, not answering it. Plants' and animals' reliance on sexual reproduction has led to many evolutionary complications (the birds and the bees! Viagra! wardrobe malfunctions!), and these make the search for answers very difficult. By studying whether *bacteria* have sex, I hope to find out why *we* do.

Bacteria don't reproduce sexually; they have one set of genes rather than two, and their cells simply divide—they never fuse with each other like eggs and sperm. But they do have some other processes that could have evolved to mix versions of genes between individuals. Short pieces of the DNA that genes are made of sometimes move from one bacterial cell to another, and this can randomly mix their different versions of genes if it happens often enough, just like real sexual reproduction does. In fact, the biologists who discovered these processes initially assumed that they evolved to do just that.

However, we now know that much of this "DNA transfer" happens by mistake. In the same way that mosquitoes can accidentally carry malaria from one person to another, tiny bacteria have even tinier parasites that can accidentally carry small pieces of DNA from one cell to another. For a long time microbiologists assumed that these parasites were a bacterial way of having sex—that is, that the evolutionary forces responsible for our sexual reproduction had indeed acted on bacteria too, producing ways to move DNA between cells. But as microbiologists learned more about these tiny parasites, they gradually realized that these didn't evolve to move the cell's DNA at all. Instead, the tiny parasites are true parasites that exploit bacteria by moving their own DNA from cell to cell (just like our viruses do). Microbiologists initially made the same assumptions about the proteins inside bacterial cells that join these DNA pieces into the cell's own chromosome (recombination), but they eventually discovered that these proteins exist to fix damage to the cell's own DNA, not to patch in foreign fragments.

There is one kind of DNA transfer that is caused not by tiny parasites but by the cell's own genes. These genes enable the bacterial cells to actively import DNA molecules from their surroundings. My research includes the study of this DNA uptake process to find out whether it exists for the same reasons as sexual reproduction.

Most microbiologists still think that bacteria take up DNA to get new versions of genes, but my research suggests that they're mistaken and that bacteria really take up DNA as food. We usually think of DNA as a set of instructions rather than a physical molecule, just as we think of a book as a story rather than layers of wood fiber and ink. And DNA is so important as genetic information that we don't usually think of it as being part of our own food, but in fact it's in every nonrefined food we eat (meat, fruits, grains, and vegetables) because it's an essential part of every cell. Bacteria get much of their food from broken-down cells in their environments, and here DNA becomes more significant because it doesn't break down (the chemical stability of DNA molecules is part of the reason they make such good genes). Some environments contain a lot of DNA, especially the slimy ones where bacteria flourish. The ooze at the bottom of a pond, the gunk on an old shower curtain, the mucus that lines our digestive and respiratory tracts—all are dining tables for bacteria that can take up DNA.

How can we find out whether bacteria take up DNA as food or as genetic information? More generally, how can we find out why bacteria do *any* of the things they do? We can't ask them directly, but that wouldn't help us anyway because this is an evolutionary *why*, not an immediate *why*. In the same way, we can't find out why sex evolved by asking people why they reproduce sexually (because it feels good, because it's the only way to have kids).

But we can indirectly ask bacteria *why they evolved* to do certain things by investigating *how they decide* when to do them. Bacteria don't have minds; their decisions are made directly, when their genes and proteins interact with their environment. When a cell runs out of its favorite sugar, it "decides" to use other sugars because a sugar-sensing protein activates the part of the cell's DNA that has genes for using the other sugars. We can also run the logic in reverse: the fact that salt-sensing proteins activate the genes for the flagella that make bacteria swim suggests that the ability to swim helps bacteria escape from toxic environments. So my research has been asking how bacteria decide when to take up DNA, because understanding what information the bacteria use to make this decision will tell us when bacteria have found DNA uptake useful.

Studying how genes are regulated thus gives us a window on how natural selection acted in the past, in the bacteria's natural environment. This may seem like a problem (don't we want to know how natural selection is acting now?) but it's actually an advantage. Bacteria are so tiny that we can't watch what they do in their natural environments; instead, we can only study them by culturing them in a laboratory under carefully controlled conditions (nutrients, temperature, no other bacteria present). These conditions are usually very different from those the bacteria experience in their natural environments, partly because they're more convenient and partly because we really don't know what the natural conditions are. But studying bacteria under such unnatural conditions makes it very easy to misinterpret what they do.

So how *do* bacteria decide when to take up DNA? First, we found out that a decision is indeed needed—bacteria don't just take up DNA any time they encounter it, but turn on the genes for taking up DNA only under certain conditions. Second, most bacteria use nutritional signals, just as the DNA-equals-food explanation predicts. My laboratory studies a kind of bacterium called *Haemophilus influenzae*. It lives in respiratory tract mucus; readers with small children may have encountered *H. influenzae* as the cause of many earaches and the target of the *H. influenzae* type b (Hib) antimeningitis vaccine. We've found out that *H. influenzae* cells decide to take up DNA when they need its building blocks—they turn on the genes when proteins signal that they're running short both of the building blocks DNA is assembled from ("nucleotides") and of the sugars that could give the energy to make more nucleotides from scratch. This is exactly how we predicted the decision would be made if DNA is food, not genes.

Although many microbiologists still assume that DNA is best used as a source of new genetic information rather than as food, they're probably overvaluing the benefits of getting new versions of genes. Taking up DNA for its genetic information is a lot like buying a lottery ticket—it's very easy to overestimate your chances of winning. Just as people who read only the newspaper headlines think every lottery ticket is a big winner, a scientist who reads only the research papers about bacterial genome sequences could easily conclude that every new gene is an improvement. But most lottery tickets are losers, and most of the DNA available to bacteria has no useful information because it's from very unrelated organisms and too dissimilar to be incorporated into their own chromosomes. On the other hand, the DNA that does come from close relatives is often worse than useless: because the only way DNA becomes available is when its original owner dies, this DNA is likely to carry harmful mutations. (I once published a paper on this topic with the title "Is Sex with Dead Cells Ever Better than No Sex at All?")

What does the lack of sex in bacteria tell us about why sexual reproduction evolved in plants and animals? Evolutionary biologists still think that plants and animals succeed better if they inherit randomly mixed versions of two parents' genes than if they inherit all the genes of one parent. But bacteria do fine even though they get only one (unmixed) set of genes from their one parent. Why the difference? The explanation must be that genetic mixing is, for some reason, of much less value to bacteria than to plants and animals. Could it be that genetic mixing does bacteria no net good at all, because its rare benefits are outweighed by the more frequent harm of replacing good genes with bad? (The lack of direct evidence of this harm is not a valid counterargument because, just as lottery losers never make the papers, losers in the genetic lottery leave no descendants for us to study.)

Bacteria may have no mechanism evolved to bring about gene mixing, but that doesn't mean that mixing isn't beneficial. Bacteria do get a modest but reliable amount of gene mixing by accident, and maybe this is enough to give them the benefits that plants and animals work so hard to get from sex. In either case, the discovery that bacteria don't have sex means that a big key to the puzzle of the evolution of sex will come from finding out why bacteria need so much less mixing than plants and animals.

Microbes and Evolution: The World That Darwin Never Saw
Edited by R. Kolter and S. Maloy

doi:10.1128/9781555818470.ch20

20

Better than Sex

Harald Brüssow

It is often said that we stand on the shoulders of giants. But sometimes we should climb down from their shoulders to take a fresh look into the faces of these giants by reading their books. Personally, I read with pleasure the books of Aristotle, arguably the founder of biology. I admire the papers of Lavoisier, the pioneer of biochemistry. Yet when reading their books, I observe a patronizing attitude—I feel naively proud about how far biology has progressed since their work. My attitude towards Charles Darwin is different. Darwin refused to simply be a historical figure. Instead of relegating me to his shoulder, he walks at my side and argues with me—he is eminently present. This freshness of Darwin is striking because he was clearly handicapped by his early birth: for example, he lacked any idea about the physical basis of heredity. Paradoxically, help was at hand. In my daydreams I encourage Gregor Mendel to write a letter to Darwin reporting on his experiments with peas, which he started in 1856. I also push Johann Fuhlrott to inform Darwin about the first skeleton of a Neanderthal man, which he found in 1856 near my home town of Düsseldorf. Darwin could have incorporated these discoveries as stunning evidence for the theory he formulated in *On the Origin of Species*,

Harald Brüssow obtained his Ph.D. at the Max Planck Institute for Biochemistry in Martinsried, Germany, and has worked since 1981 at the Nestlé Research Center in Lausanne, Switzerland. He enjoys family life, working in a multicultural and multidisciplinary environment, reading classical books, and hiking in the Swiss Alps.

which was published in 1859. However, Mendel described his results in a letter to an obscure scientific journal in his home city of Brünn, Austria. And, although Fuhlrott wrote a letter to a scientific celebrity, Rudolf Virchow, the founder of molecular medicine, Virchow misinterpreted the Neanderthal man as a rickets-afflicted Cossack soldier.

In a recent review I qualified Darwin's book as the Great Unifying Theory of biology. Physicists are still dreaming of such a theory explaining what holds the entire physical world together. Biologists have had such a theory for the last 150 years. Well, his theory needed a few updates to incorporate genetics and molecular biology, but the ease with which this theory incorporated these novelties showed its incredible strength. Scientists, particularly microbiologists, continue to expand on the theory. The single picture in Darwin's book displays a hypothetical tree of organisms. Microbiologists helped to put flesh on this skeleton by formulating the universal tree of life based on genome sequence information. Darwin's early notebook sketch of a tree could come from a modern textbook. It was again microbiologists who modified the tree analogy by discovering the importance of horizontal gene transfer (i.e., transmission between genetically unrelated organisms). They replaced the tree with a network of criss-crossing interconnections. We do not yet know the solution for this thorny problem—trees have become thickets—but as set forth by Greek philosophers, from thesis over antithesis a new synthesis grows. Darwin's ideas have continually shown their eminent creativity and capacity to integrate seemingly conflicting new observations. The theory from yesterday becomes a special case of a broader theory of tomorrow, which is the best criterion for the value of a scientific theory. Science is, anyway, not about beliefs written in stone—beliefs are inherently static, backwards-oriented, and conservative and thus prone to be outdated—but about a method to get to new insights.

What are current problems with Darwin's theory? In a recent book, I reasoned that the fundamental driving forces of organisms are the quest for food, the avoidance of predation, and the quest for sex. The only problem with this statement is that bacteria have no sex—what microbiologists call mating in bacteria is a far shot from real sex—and this fact has indeed far-reaching consequences. Chapter VIII of *On the Origin of Species* links sexuality with the species concept. This is a time-honored, although not very practical, biological species definition that serves as a theoretical basis for much evolutionary reasoning. If you have no sex, you do not qualify as a biological species. In their daily practice, microbiologists use species names in the binomial tradition of Linnaeus because they need names for observable different entities in their part of the biological world. A name means that we can set these organisms apart from other types of organisms. Even if you forget for the moment the problem with the theoretical underpinnings of the bacterial species concept, there are also practical problems. Currently we have fewer than 10,000 named bacterial species. Not to mention the notorious beetles, 10,000 is just the number of the described species of mites (*Acarina*). Of course, this number is a gross underestimate. As science is about numbers, the fact that we do not have a well-based estimate for the number of bacterial species is revealing. The joke that the best estimate is obtained by asking experts about their personal guess, then applying a statistical model on these guesses, is not so far off the mark. As a biological species definition cannot be applied in bacteriology, bacterial species are defined by a bundle of criteria, including DNA sequence diversity. The DNA sequence diversity even within well-accepted bacterial "species" is enormous: independent isolates frequently differ by 10%, compared with the difference of 1 in 10,000 bp among human beings and the 1% difference between humans and chimpanzees.

From first principles, it is not clear why sexless bacteria should be so diverse. Sex is commonly interpreted as the motor for genetic diversity in a population. In the bacterial world a male germ cell does not fuse with a female germ cell to create a new daughter cell. Instead, a "parental" bacterial cell grows in size and then divides by a fission process into two daughter cells. In fact, logically it makes no sense to speak of parental and daughter cells—at the very moment when you have two daughter cells there is physically no parental cell left. Fission has a genetic consequence: parental and daughter cells have identical sequences. A priori, there is no genetic individuality in bacterial lineages; all member bacteria are "dividuals." Therefore, you would expect that bacteria are strictly clonal organisms when neglecting rare spontaneous mutations. This conclusion has interesting consequences for cooperation and altruistic suicide in bacterial populations. Furthermore, sexless bacteria are in principle immortal—this link between sex and mortality was already intuitively felt in the book of Genesis (nota bene: an interesting reading for biologists, regrettably obscured by the creationist discussion). So from these philosophical considerations, bacteria should be the most static organisms in biology, but they clearly aren't. On the contrary, bacteria evolve within decades, much more quickly than higher organisms. Apparently, bacteria have found a supermotor for diversification, which is much more efficient and which works more quickly than sex.

My personal experience with this bacterial conundrum goes back to my research in dairy microbiology. Viral infections of bacterial starter cultures are a major problem in industrial milk fermentation. When we studied the genomes of bacteriophages (bacterial viruses) and dairy bacteria to find an Achilles' heel for an intervention, we were surprised by the poor defense that the yogurt starter *Streptococcus thermophilus* mounted against phage attack. CRISPRs—DNA repeats in the starter providing a form of acquired genetic immunity against phages—were not yet known at that time. We even observed molecular evidence that bacteria cooperate in the integration of phage DNA into their chromosomes (prophages), in contrast to textbook descriptions of phages and bacteria as enemies locked in an age-old arms race. When we later collaborated with clinical microbiologists, we observed that prophages from pathogenic streptococci carried genes that increased the disease potential of the bacteria. When prophages become part of the bacterial genome, selection works on the prophage-bacterium consortium. Prophages apparently "helped" *Streptococcus pyogenes* become a successful pathogen, and over a few decades *S. pyogenes* changed its phenotype. Some strains cause streptococcal angina; others are associated with skin infections, scarlet fever, rheumatic fever, or kidney diseases. The latest phenotype is what the press calls flesh-eating bacteria. Despite these dramatic phenotypic changes, the bacterial chromosome has practically not changed. What has changed, however, is the prophage content of the pathogen. Since *S. pyogenes* contains multiple prophages, each coming with its own virulence genes, bacterial pathogenicity was apparently created by the combinatorial principle of distinct prophage assortments. Phage provide a mobile source of new DNA, and bacteria can rapidly screen the viral DNA sequence space for useful genes. This process is much quicker than evolving new pathogenic traits de novo. Phages are thus not simply an enemy of the bacterial cell but also a motor of bacterial evolution. This new relationship can be rationalized with basic Darwinian reasoning. At the quest-for-food level, bacteria are the prey not only of phages but also of single-celled higher organisms called protists. In addition, protists can become food for those bacteria, which succeed in killing them by escaping the protist's food vacuole. Since we use macrophages (cells that strikingly resemble amoebae, prominent protists that prey on bacteria) to defend our body against bacterial invaders, pathogenicity becomes a special case of the battle for food. Bacterial pathogenicity can thus be understood from the rules of the rock-paper-scissors game when using phage-bacteria-protists in this argument. It is revealing that the medical name macrophage—big eater—semantically reflects the quest-for-food argument, as does bacteriophage—bacterium eater. You might not be surprised that a scientist working in the food industry would argue for the priority of eating over sex in evolution.

S. pyogenes is not an isolated case: many important human pathogens owe crucial parts of their pathogenic potential to bacteriophage-carried genes. Prophage DNA is also very prevalent in nonpathogenic bacteria. Without exaggeration one might argue that phages—together with other mobile DNA like plasmids or transposons—are the major motor for the generation of genetic diversity in bacteria. Bacteriophages provide a type of infectious sex for bacteria. However, as infections are typically horizontal events occurring between unrelated individuals—in contrast to sex, which relies on vertical gene transmission, i.e., from parent to progeny—the problem with the complicated structure of the bacterial tree finds an easy explanation.

Our reflections are rather typical for Darwin's theory. Evolution depends on selection applied to genetically distinct organisms, but whether this diversity is created by sex or bacteriophage infection does not matter. Sex is a special case of a more generally defined driver of biological diversity. In fact, Darwinian evolution was operative long before sex evolved.

At the end I should mention another idiosyncrasy of the theory. The discussion about the tree of life is not only about its branching pattern, but also is now about the tree itself. After extensive sequencing of viral genomes, it has become increasingly clear that our "universal" tree of life is only a tree of cellular life. Noncellular biological entities like viruses do not belong to this tree. Viral genomes apparently do not derive from cellular genomes; they must come from a different source. Will this kill the concept of a universal tree? I guess so. Will this affect the Darwinian theory? Probably, but I suppose that this finding can also be integrated into another, updated Darwinian theory. We progress thus from a "special" to an increasingly more "general" theory of evolution. In fact, some people claim that Darwin's concept of evolution applies also to nonbiological systems. You might now understand why I want to walk at the side of this great biologist instead of standing on his shoulder. He might still have a lot up his sleeve, and in my daydreams I still continue to encourage people to write letters about their latest discoveries to him.

FURTHER READING

Brüssow H. 2007. *The Quest for Food: A Natural History of Eating*. Springer, New York, NY.

Microbes and Evolution: The World That Darwin Never Saw
Edited by R. Kolter and S. Maloy

doi:10.1128/9781555818470.ch21

21

Darwin in My Lab
Mutation, Recombination, and Speciation

Miroslav Radman

Here is a theatrical Gedankenevent stimulated by thinking about Charles Darwin. A resurrected Charles Darwin walks into my lab and, without humor, says, "You mention my name too often, Radman. Now, do you have anything really interesting to show me, keeping in mind that I have learned absolutely nothing since my last publication and that I prefer seeing things before making any conclusions?" I stutter, "With all due respect, but if you had previously Googled Gregor Mendel, Hermann Muller…."

Luckily, I believe that I could offer him a true Darwinian treat. Namely, the resurrected Darwin could watch with us movies revealing the two kinds of molecular processes that have generated all diversity of life on Earth—mutation and recombination. We could also observe directly and in real time individual mutation and recombination events emerging spontaneously in bacterial cells as they grow under the microscope objective. By counting each fluorescent flash, diagnostic of a new mutation, and each breakage of fluorescing DNA, diagnostic of a crossover, we can monitor all of

Miroslav Radman trained at the universities of Zagreb, Brussels (Ph.D.), Paris, and Harvard, and at age 28 became the youngest professor at Brussels University. In 1983 he moved to Paris, where he is Professor at R. Descartes University Medical School and a member of the French Academy of Sciences. He is a bon vivant madly in love with science and the arts.

the individual mutation and recombination events in any lineage of *Escherichia coli* cells, independently of their potential phenotypic effects.

After the more than 150 years since *On the Origin of Species* was published, we can directly observe the two kinds of processes that create heritable variation: (i) "vertical" variability, by continuous emergence of new point mutations that keep adding to older ones, and (ii) "horizontal" variability, by new arrangements of already existing genetic variants. The phylogenetic trees (like the tree drawn by Darwin in *Origin*) are the result of vertical variation, whereas incongruencies in such trees are a consequence of horizontal gene transfer via genetic recombination between nonidentical parental genomes. Although mixing of parental heritable characters in their offspring was familiar to Darwin, explaining to him about DNA, how genes produce phenotypes, and how horizontal gene transfer occurs would take some time. If Darwin's visit coincided with Matt Meselson's 1-month annual *séjour* in our lab, I could rest for a while and enjoy listening to Matt explain to Charles how DNA replicates and recombines.

Our methods for direct visualization of individual mutation and recombination events in living *E. coli* bacteria are based only on the shrewd use of a natural DNA modification (methylation of A in the GATC sequences) and of two fluorescent proteins. A special protein involved in the process of DNA error correction (called mismatch repair) binds exclusively, extensively, and stably to uncorrected DNA copy errors, forming a fluorescent focus in the living cell. When, in the next DNA replication round, the copy error itself is copied and converted to a permanent mutation, the focus vanishes. Thus, we can visualize the time course of emergence and fixation of each new mutation!

Now, let us watch recombination. A fluorescent version of the *E. coli* SeqA protein, which binds exclusively the hemimethylated DNA (a DNA duplex with only one strand methylated), allows us to monitor the integrity of permanently hemimethylated DNA over an unlimited number of generations. Such stably hemimethylated DNA is created when a methylation-deficient bacterial mutant cell receives and integrates a methylated DNA strand from a nonmutant DNA donor. Such methylation-deficient cells will retain the methylated DNA strand in its integrity for any number of cell generations unless a crossing-over breaks it, followed by segregation of one fragment with another unmethylated chromosome. That is exactly what we see occurring only in recombination-proficient cells, hence providing a method for direct visual monitoring of crossovers.

The rates we calculate for spontaneous mutation and recombination events (0.004 and 1.4 per cell, respectively) in the bacterium *E. coli* are as follows.

1. 1.4×10^{-8} per base pair is the error rate of the DNA replication machinery, although this rate is reduced by the mismatch repair system to the final genomic mutation rate of 1.8×10^{-10} mutations per base pair, per cell generation
2. 3×10^{-7} crossovers per base pair, per generation

These figures correspond closely to the mutation and recombination rates per genome extrapolated from the frequencies of single locus events by geneticists.

The molecular phylogenies of preexisting mutations from genomes of many sequenced natural isolates of *E. coli* suggest that the acquisition of a mutation at any locus in the genome is about 100 times more likely to occur by recombination than by de novo mutation. Furthermore, analysis of many genomes of natural *E. coli* isolates shows that they all share only about 2,000 genes, whereas 18,000 different genes are present only in the many different strains of *E. coli*. This gives us an idea how flexible, fuzzy, and dynamic the genetic identity of bacteria is when they indulge in frequent horizontal gene transfer.

Now, if DNA transfer and recombination were very promiscuous and frequent, there would be no distinct species and therefore the actual biological diversity would be low: there would be one "bastard" species. In order to create distinct genetic diversity in very large numbers of organisms, there must be barriers to the free gene flow among populations (i.e., species). Such interspecies barriers exist and are used to define species according to Ernst Meyer (his term "reproductive isolation" refers really to the absence of frequent gene mixing). Populations unable to share their genes—even when put in close contact—are considered different species. These genetic barriers are of many kinds (it is pretty evident why a bird and an elephant cannot easily interbreed), but most puzzling are those that do not preclude either the mating act itself or the physical proximity of two genomes within the same cell, but involve interdiction for parental genomes to recombine and give fertile progeny. These are genetic barriers between closely related species. The reproductive sterility of sturdy interspecies hybrids, mule and hine, defines horse and donkey as different species.

Our studies of enterobacteria provided detailed insight into the molecular nature of genetic barriers between related species, such that these genetic barriers became subject to easy manipulations: they can be either largely eliminated or greatly amplified, simply by controlling the amount of two specific proteins that prevent interspecies recombination. Proteins that prevent pairing and recombination between nonidentical DNA sequences are the key mismatch repair proteins, MutS and MutL, which are conserved throughout the kingdoms of life. MutS recognizes and binds to all mispaired and unpaired bases in DNA (except C:C mismatches), whereas MutL binds exclusively to the mismatch-bound MutS and establishes communication with other mismatch repair proteins such as UvrD (helicase II) and MutH (which cuts the newly synthesized strand at an unmodified GATC sequence).

Here is our current image of the reversibility, or abortion, of the initial stage of recombination when the two partner DNA molecules (or sequence blocks repeated within the same DNA molecule) are diverged in their nucleotide sequence, even at a very low level. First, the broken end of one molecule is exonucleolytically processed, liberating a free 3′ single-stranded tail that is coated by the key recombinase, RecA, that catalyzes its insertion into the homologous sequence of a DNA duplex, forming a three-stranded "D-loop." When the nucleotide sequences of the two partner molecules are nonidentical, the D-loop heteroduplex region will contain mispaired or unpaired bases (mismatches). Such mismatches are readily recognized and bound by the MutS protein, which attracts the MutL protein, which in turn will attract UvrD (helicase II), melting the DNA heteroduplex from the invading 3′

end in the 3′-to-5′ direction—undoing thereby what was achieved by the RecA recombinase. In this way, the attempt to recombine nonidentical DNA sequences is aborted. Interestingly, the amount of MutL is limiting in this antirecombination activity of the MutS, MutL, and UvrD complex such that overexpressing MutL creates genetic barriers even among *E. coli* strains that have exhibited unhindered genetic recombination. Null *mutS* or *mutL* mutants show a great promiscuity in that even genomes diverged by 20% (e.g., *Escherichia* and *Salmonella* species that have been diverging for 100 million years!) can efficiently recombine and replace up to one-third of each other's genome. We used to call such interspecies hybrids "*Salmorichia*" (when the resulting genome was mostly *Escherichia*) or "*Eschenella*" (when the resulting genome was mostly *Salmonella*).

A subsequent demonstration that the meiotic sterility of yeast interspecies hybrids between *Saccharomyces cerevisiae* and *Saccharomyces paradoxus* (10% genomic sequence divergence) can be largely alleviated by mutations in the *mutS* and *mutL* homolog genes (*MSH2* and *MLH1*) was the vindication of our bacterial model of genetic barriers in the world of eukaryotes.

By now Darwin would have been hopefully excited, but surely exhausted in spite of my using a lot of drawings. At this point Darwin might have said, for example, "Well, I asked your student to Google H. Muller for me and I learned that something called radiation causes mutations. If you are so sure of your speciation model, could you perhaps heavily mutagenize your *E. coli* over and over, and thereby initiate speciation in your lab even before you die of old age? If you could create new species by genetic variation within a human lifetime, what else could I expect from my resurrection?" I would respond triumphantly, "We already did it, Charles, but more naturally and elegantly—without use of radiation, but with the cunning use of Richard Lenski. You must visit this smart fellow at the University of East Lansing, who did a most boring and brilliant experiment that brought him a score of interesting papers (including one with us). Let me tell you about it...."

In 1997, after over 12 years of a daily regimen of dilution and regrowth of bacteria in a large series of cultures, Lenski and colleagues realized that 4 out of 12 parallel cultures (derived from the same initial *E. coli* clone) had become mismatch repair-deficient *mutS* or *mutL* "mutators" after about 1,000 generations. This interesting and counterintuitive observation had been predicted by modeling in our lab. What mattered for our "speciation experiment" was that in those four cultures that had become mutators, the cells had grown for up to 18,000 generations with mutation rates about 100 times higher than those of the ancestral strain. Thus, evolution should have been greatly sped up! We calculated that about 0.2% sequence divergence should have accumulated between genomes of two independently growing mutator cultures. Rich Lenski generously provided what we needed: the ancestral strain and two independently derived mutator and nonmutator cultures. Marin Vulic first restored by recombination (transduction) the functional *mutS* and *mutL* genes in the defective mutator strains. Now, he could cross different strains—all derived from a single clone—and all with either functional or nonfunctional mismatch repair, differing only in whether they promoted either high or normal mutation rates.

The results appeared as expected from our *E.coli–Salmonella enterica* serovar Typhimurium crosses: when both partners had mutator histories, genetic recombination dropped 10-fold in *mut+* cells but not in cells with *mut* defects; when only one partner had mutator history, recombination dropped only 3- to 4-fold; but when neither had a mutator history, there was no effect on recombination. Thus, the mutator regimen produced genetic polymorphism that would require 100 times more generations for the wild-type partners. Clearly, the accumulation of mutator mutations buys a lot of evolutionary time! After only 12 years we could see the buildup of the genetic barrier—i.e., an emerging speciation event! In other words, the genetic polymorphism is the "substrate" and the MutS/L proteins are the "enzymes" of sympatric speciation. We wrote a nice paper with Rich Lenski entitled "Mutation, Recombination, and Incipient Speciation of Bacteria in the Laboratory" that was published in the *Proceedings of the National Academy of Sciences of the United States of America*. I expected a splash in the community of evolutionary geneticists, but there was—and remained—silence! This paper is in the good company of the best paper I ever published and shares with it the lowest citation record. Well, I still like it, and I bet that the resurrected Darwin would share my taste for homemade evolution.

FURTHER READING

Babic A, Lindner AB, Vulic M, Stewart EJ, Radman M. 2008. Direct visualization of horizontal gene transfer. *Science* **319:**1533–1536.

Drake JW, Charlesworth B, Charlesworth D, Crow JF. 1998. Rates of spontaneous mutation. *Genetics* **148:**1667–1686.

Elez M, Murray AW, Bi LJ, Zhang XE, Matic I, Radman M. 2010. Seeing mutations in living cells. *Curr Biol* **20:**1432–1437.

Elez M, Radman M, Matic I. 2007. The frequency and structure of recombinant products is determined by the cellular level of MutL. *Proc Natl Acad Sci USA* **104:**8935–8940.

Guttman DS, Dykhuizen DE. 1944. Clonal divergence in *Escherichia coli* as a result of recombination, not mutation. *Science* **266:**1380–1383.

Hunter N, Chambers SR, Louis EJ, Borts RH. 1996. The mismatch repair system contributes to meiotic sterility in an interspecific yeast hybrid. *EMBO J* **15:**1726–1733.

Jones M, Wagner R, Radman M. 1987. Mismatch repair and recombination in *E. coli*. *Cell* **50:**621–626.

Rayssiguier C, Thaler DS, Radman M. 1989. The barrier to recombination between *Escherichia coli* and *Salmonella typhimurium* is disrupted in mismatch-repair mutants. *Nature* **342:**396–401.

Taddei F, Radman M, Maynard Smith J, Toupance B, Gouyon PH, Godelle B. 1997. Role of mutator alleles in adaptive evolution. *Nature* **387:**700–702.

Vulic M, Dionisio F, Taddei F, Radman M. 1997. Molecular keys to speciation: DNA polymorphism and the control of genetic exchange in enterobacteria. *Proc Natl Acad Sci USA* **94:**9763–9767.

Vulic M, Lenski RE, Radman M. 1999. Mutation, recombination, and incipient speciation of bacteria in the laboratory. *Proc Natl Acad Sci USA* **96:**7348–7351.

Microbes and Evolution: The World That Darwin Never Saw
Edited by R. Kolter and S. Maloy

doi:10.1128/9781555818470.ch22

22

Sexual Difficulties

Howard Ochman

Despite those pleasures (or anxieties) that we associate with sex, its true biological purpose is to transfer heritable genetic material between organisms. But sex in bacteria is very unlike the fecundations of plants and animals. Bacterial sex does not require the formation or fusion of gametes, and there is often no recombination or creation of new combinations of chromosomal genes. Moreover, bacterial sex need not involve cell-to-cell contact or occur between members of the same species. Most importantly, sex in bacteria is not linked to reproduction—parents passing genes to offspring (via that cycle of sex and reproduction) is the opposite of what we consider to be sex in bacteria. In bacteria, sex is the inheritance of DNA from any source *except* the parental cell.

Bacteria reproduce asexually, whereby a cell replicates its chromosome and splits into two daughter cells that each inherit identical copies of the parental genes and chromosome. Mutations are rare—there is on the order of only one error per 1,000 rounds of replication—so most cells are clones of their ancestors, siblings, and descendants. At first glance, this low input of genetic variation

Howard Ochman was trained at the University of Rochester and UC Berkeley, has held appointments at Washington University, the University of Rochester, the University of Arizona, and Yale University, and has been lucky enough to work with (and, in one case, marry) the most fun, interesting, and inspiring collaborators during his academic career.

seems hardly sufficient to advance the diversification or adaptation of any life form, but remember that bacteria have four things going for them.

The first is that asexual reproduction results in an exponential increase in cell numbers: it takes 10 generations to attain 1,000 cell replications and only 70 generations to produce 10^{21} cells, equal to the estimated number of stars in the universe. Second, the generation times of bacteria are short, on the order of hours or days for those species considered to be "slow growing." Therefore, relatively little time is needed for cell numbers to reach the point where each nucleotide in a genome has been subjected to at least one mutation. The third feature is that bacteria are very small (and seem not to mind crowds or living with relatives). At the size of bacteria, all of the inhabitants of Tokyo could fit in a single drop. This means that your glass of pasteurized milk—recognizing that pasteurization eliminates 99.999% of all microorganisms—still contains a number of bacterial cells equivalent to the population of Mexico City. So despite the extremely low input of mutations and genetic variation during the process of asexual reproduction, the standing populations of bacteria are usually so large that there are ample variants ready to adapt to some new environmental circumstance.

And then there is the fourth factor that contributes to the adaptation and diversification of bacteria: sex. Bacterial sex creates variants of a very different sort from those produced during replication/reproduction. Whereas the majority of changes that occur while DNA is replicating involve the alteration of only a single nucleotide within the entire genome, sex can transform vast regions or introduce hundreds of new and exotic genes. Therefore, the variants generated though the sexual acquisition of DNA are capable of behaving in ways that are very different from their immediate ancestors and completely novel for the species.

The particular type and amount of DNA that bacteria might gain through sex depend on both its source and the mechanism by which the DNA is procured. The transfer of genetic material can occur by any of several processes which were each characterized decades ago. Conjugation is the closest to what we might normally consider "sex," since it entails the transfer of DNA from a donor cell to a recipient joined by a slender appendage, analogously termed the pilus (Fig. 1).

When this occurs between members of the same species, the transferred genes are often similar in function, but not necessarily in DNA sequence, to those that already exist in the recipient genome; consequently, exchanging them with the resident genes will modify the sequence of large regions of the recipient genome. In the more promiscuous cases, when cell-to-cell contact and DNA transfer occur between very divergent organisms, the recipient can

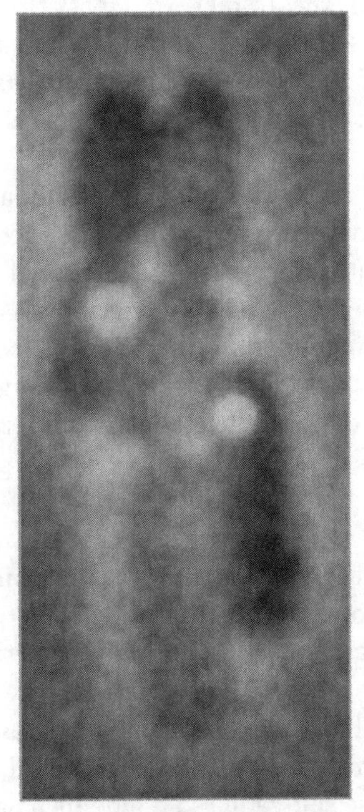

Figure 1 Conjugation in bacteria and that in humans both involve physical contact between the DNA donor (central cells in bacteria; tall fellow in humans) and a recipient. The round afterglow spot observed in the bacterial recipients—outer cells—denotes a successful DNA transfer event. Micrograph courtesy of Ana Babic, INSERM. doi:10.1128/9781555818470.ch22f1

gain some unique genes (and traits) that were previously restricted to the donor.

Bacterial sex, which is also termed horizontal (or lateral) gene transfer due to the fact that the genetic material is being passed across, and not vertically down, lineages, also results from processes known as transduction and transformation. Bacterial viruses, known as bacteriophages or simply as "phage," which make their living by inserting DNA into bacteria, serve as the DNA transfer vehicle for transduction. Because phage can assume large multigene fragments and infect a wide range of hosts, they appear to be the primary source of novel genes in bacterial genomes by facilitating the transfer of genes across broad phylogenetic distances. Finally, transformation is the uptake of DNA from the environment. Often, this process involves the replacement or incorporation of short patches of the host genome, but it has also been implicated in the acquisition of larger DNA elements originating from other genomes.

It would appear that the major benefit of sex in bacteria is to offer a rapid means by which new properties are added to the bacterial cell. The wholesale acquisition of fully functional units that can be deployed immediately upon arrival seems to be a much more efficient process than the incremental alteration of existing genes when trying to generate a new and needed trait. And not surprisingly, several incidences of rapid bacterial evolution and adaptation, such as antibiotic resistance and pathogen emergence, have been instigated by events of gene acquisition.

There is an alternative view, which poses that any advantage tendered by acquired DNA is secondary to the real reason that bacterial sex evolved. Suppose, for example, that individual genes each have the goal of propagating themselves and that bacteria are merely the vessels by which this can occur. Therefore, those alien sequences (think of the movie "Alien") that occupy (i.e., parasitize) the broadest range of host genomes are the most successful. And any useful traits provided by these self-centered elements emerged simply to bolster the process, since by benefitting their hosts, genes ensure their own maintenance and survival.

Whatever their origins or purpose, the existence of bacterial sex and gene transfer has provoked both confusion and contention among biologists. Because bacteria lack the overt traits that classical systematists use to describe and classify multicellular organisms, the evolutionary relationships among bacteria are based largely on their gene sequences. The reasoning is as follows: if two bacteria derive very recently from a common ancestor, the DNA sequences of their corresponding genes will be the same. And since mutations accumulate through time, distantly related organisms will have dissimilar sequences, and in fact, the degree of dissimilarity should directly reflect the evolutionary distance between the organisms.

But introduce sex into the equation, and this logic becomes somewhat hazy. If a gene is acquired from a divergent source, both the donor and recipient will have identical copies despite the fact that the organisms themselves are unrelated. Expanding this to the point where every gene has the potential to transfer to any organism from any source, and you see why this might cause some systematists to lose sleep: rampant gene transfer would render useless their efforts to assign microbes to meaningful groups and to reconstruct the evolutionary links between organisms, since the closest relatives may be no longer be the most closely related. This defies the way that Darwin and subsequent naturalists, geneticists, and evolutionists define and view the biological world.

Whereas it is broadly acknowledged that bacterial genomes are mosaics of genes derived from both ancestral and distant sources, the utility of genes for determining the evolutionary relationships among bacteria has been a tortured topic. Some have considered it hopeless, saying that we should abandon attempts at phylogenetic classification because each organism might be perceived as belonging to several taxonomic groups. But we need to remember that there is a continuous cellular lineage that traces from every contemporary organism back to our universal common ancestor. And because this is the case, there is beyond doubt a Tree of Life connecting all organisms, and it is the task of biologists to find ways to resolve, depict, and understand it.

Microbes and Evolution: The World That Darwin Never Saw
Edited by R. Kolter and S. Maloy

doi:10.1128/9781555818470.ch23

23

Unveiling *Prochlorococcus*
The Life and Times of the Ocean's Smallest Photosynthetic Cell

Sallie W. Chisholm

> Our task now is to resynthesize biology; put the organism back into its environment; connect it again to its evolutionary past; and let us feel that complex flow that is organism, evolution, and environment united.
>
> Carl Woese, 2004

A few years ago, the woman who cuts my hair—who has become a part of the fabric of my life over the years—finally worked up the courage to ask me about my research. I explained that we study a very small microorganism that lives in the surface oceans—a tiny single-celled plant, discovered only 25 years ago. "You've been studying the same organism for 25 years? That must get *really* tedious!" she said as she snipped away at my hair for the 100th time.

It is difficult to describe the thrill of studying *Prochlorococcus*. The name alone is enough to stop a conversation. Far from being tedious, studying this extraordinary little cell is like opening a present every day. It is a gift, and a responsibility. When people ask

Sallie W. ("Penny") Chisholm obtained her Ph.D. from S.U.N.Y. Albany and did postdoctoral research at Scripps Institution of Oceanography. She joined the MIT faculty in 1976 and has been there ever since. When not pondering *Prochlorococcus*, hanging around in Woods Hole, and learning from her students, she enjoys dipping her toes in Lake Superior and skiing in Colorado.

me about it I usually launch into my "photosynthesis appreciation" lecture, trying to convince them that nearly all life on Earth comes from photosynthesis: making life from sunlight, air, and water. If I succeed at that, which is not easy, I go on to tell them that half of global photosynthesis is done by microscopic phytoplankton in the oceans and that *Prochlorococcus* is the smallest and most abundant member of this "invisible forest." There are over a trillion trillion *Prochlorococcus* organisms in the global oceans, and we did not know that they existed until a few decades ago. People find this extremely difficult to believe. I would too, had I not been lucky enough to live through its story.

Like most scientific advances, the unveiling of *Prochlorococcus* involved new technologies, diverse approaches, teamwork, and luck. Until about 40 years ago, we thought that all phytoplankton were between 5 to 100 μm in diameter, because this was all we could easily see under a microscope. (I use the editorial "we" throughout this essay for simplicity, but it refers to many people involved in this story; see last page for elaboration.) In the 1970s, advances in microscopy revealed that the oceans were filled with even smaller photosynthetic cells, about 1 μm in diameter, that were 10 times more abundant than the larger phytoplankton. Because of their unusual pigments, these cells, ultimately named *Synechococcus*, appeared as tiny orange beacons. Higher-resolution images of their populations revealed subtly different variants, and one of the smaller variants was more abundant in deep water and appeared more green than orange on collection filters. This was *Prochlorococcus*. But as is often the case with major discoveries, its photo sat unnoticed on the journal page for a decade. Around that same time, Dutch oceanographers discovered an "unidentified chlorophyll derivative" in samples from the North Sea that was particularly abundant in particles smaller than 1 μm. This was, in fact, a pigment that we now know is uniquely characteristic of *Prochlorococcus*, but they had no way of knowing that at the time.

Over a decade later, the pieces of the *Prochlorococcus* puzzle began to fall into place. An instrument became available to oceanographers that employs lasers to study the pigment and light scatter properties of microbes in seawater. It was ideal for studying *Synechococcus* because of its distinct orange pigmentation, so study it we did. After several years, we began to notice some signals emerging from the electronic noise of the instrument that suggested the presence of cells that were not orange but green. We ignored them for a while, thinking they were just an extension of the noise signal. But eventually we noticed that the signals were stronger in samples from deeper water, where a cell might need more pigment to harvest the dwindling sunlight. This could not be ignored. It soon became clear that there were photosynthetic microbes smaller than *Synechococcus* in the oceans, with a different pigment, that were 10 times more abundant. There were 100 million of these cells in a liter of seawater.

Over the next few years, we determined that these were the very cells that had been photographed a decade earlier and dubbed *Synechococcus* variants. And we learned that they contained divinyl chlorophyll, which has properties identical to the "chlorophyll derivative" that Dutch oceanographers had described passing through their 1-μm-pore-size filters years before. We discovered further that the cells also contained chlorophyll *b*, a pigment that is typically found in "green plants" found on land. With growing affection, we began to call our newly discovered cells "little greens." Their detailed structure bore an uncanny resemblance to chloroplasts, the small oval bodies in plant cells where photosynthesis takes place, which are known to be evolutionarily derived from microbial cells through an ancient symbiotic union. No microbe had been identified that both resembled a chloroplast and contained the telltale chlorophyll *b*. Had we found the missing link? Were our little greens living fossils? Before we knew the answer to these questions, we had to give the cells a proper name. We chose *Prochlorococcus*—"little round progenitors of chloroplasts." We would soon regret this.

Around that time, a revolution was occurring in evolutionary biology. Scientists had learned that the relatedness among organisms could be measured by comparing the DNA sequences of genes that are shared universally across all living things. By sequencing these genes in *Prochlorococcus*, we could ask directly, does *Prochlorococcus* share a recent common ancestor with the chloroplasts of higher plants? The answer was no. *Prochlorococcus*'s closest relative is, perhaps not surprisingly, *Synechococcus*, and neither of them shares a recent common ancestor with chloroplasts. We had come full circle. *Prochlorococcus* is just a smaller version of *Synechococcus*, with an unusual set of pigments. But once a microbe has a name, it takes a lot of effort to change it. *Prochlorococcus* stuck.

The first *Prochlorococcus* cells isolated into culture came from a 120-m depth in the Sargasso Sea. Soon thereafter, another was isolated from the surface waters of the Mediterranean Sea, and the strains were named SS120 and MED4, respectively. It became clear immediately that although they shared the "signature" characteristics of *Prochlorococcus*, MED4 and SS120 were not the same: MED4 could grow at high light intensities that killed SS120, while SS120 could grow under extremely low-light conditions that could not sustain MED4; i.e., the cells were adapted to the light intensities found where they were captured. We called them high- and low-light-adapted "ecotypes." It was slowly dawning on us that it was through the layering of these ecotypes that *Prochlorococcus*—the collective—was able to fill the sunlit 200 m of the oceans. We had captured two of them. How many more were there? Did they differ in other ways?

Over the years that followed, we isolated more strains of *Prochlorococcus* from many different oceans and depths. We sequenced their diagnostic genes and used them to develop a family tree of sorts. The strains could be grouped into two broad clusters—either high- or low-light adapted. Within the high-light-adapted group there were two additional clusters. What differentiated them? We went off to sea and mapped their distributions along a north to south transect of the Atlantic Ocean, letting the cells tell us, through their relative abundances, what environments most suited them. The answer was clear. One ecotype dominated the warm tropical and subtropical waters, and the other the colder waters at high latitudes. Further experiments showed that strains differed not only in their temperature preferences but also in their abilities to exploit different nitrogen and phosphorus sources. Thus, *Prochlorococcus* is a federation of sorts, divvying up the oceans according to light, temperature, and nutrients. What else? How many different evolutionary paths have these cells taken?

While we were growing cultures and mucking about at sea, the DNA sequence of the entire human genome, containing roughly

20,000 genes, was completed and released with much fanfare. This left a lot of DNA sequencing machines available for other projects, and because of its small size, photosynthesizing ability, and global reach, *Prochlorococcus* MED4 was among the first microbes to have its genome completely sequenced. While I was very excited about this opportunity, I could not shake the absurd feeling that we were invading the inner life of this tiny cell that had drifted unnoticed in the oceans for millions and millions of years. My feeling grew into a sense of responsibility—a need to bring it the respect it deserves. One does get attached.

The first thing we learned from the genome sequencing project was that MED4 is very streamlined, even for a microorganism, containing only about 1,700 genes. So far, this represents the minimum amount of information (DNA is simply information in chemical form) necessary to create life out of elemental components: sunlight, carbon dioxide, water, and other essential nutrients drawn from seawater. This cell is truly the "essence" of life. As a photosynthesizer, it can do what humans cannot, even with all of our technology: it can split water using sunlight and make hydrogen and oxygen—all with only 1,700 genes.

In studying evolution, it is the difference between things that holds the answers, so we sequenced the genomes of 12 *Prochlorococcus* strains spanning the family tree. There are 1,200 genes shared by all of them. This is its essential core—the bare bones of being *Prochlorococcus*. Each strain is endowed with 500 to 1,200 additional genes, some shared with some (but not all) of their cousins. These non-core genes give each strain its unique character. Some determine, for example, what nutrients the cells can scavenge, how well they can protect themselves from high light, and what their outer surface looks like to predators. But the functions of most of these extra genes are a total mystery to us. They hold important keys to understanding the sheer abundance and persistence of *Prochlorococcus* in the oceans and have much to tell us about the selective forces that have shaped both the ancient and recent oceans. We think of the cells as little "reporters" who use a language we only partially understand.

The sheer number of these reporters and the genes they carry is astounding. So far, every new strain that has been sequenced has revealed an average of 200 completely new genes. Thus, the size of the global *Prochlorococcus* federation gene pool—the so-called pan genome—must be enormous. Is each cell unique, like a snowflake? We know this cannot be true, because the cells reproduce by making identical daughter cells. Each cell grows by day, basking in the sun's energy, and each divides into two by night. They all do this in unison, coordinated by the daily pulse of energy. For every cell that is produced, there is another that is eaten by small predatory cells that must rely on others for their food. This keeps the *Prochlorococcus* population in check and begins the flow of energy through the marine food web.

Because they reproduce by making identical copies of themselves, we know that at any moment in time there must be lineages of identical *Prochlorococcus* cells in the oceans. But how many lineages are there? What is the rate of genetic change and how does it occur? While we don't have answers to these questions, we are finding clues where we might least expect them: viruses. There are millions of viruses in a milliliter of ocean water, some of which use *Prochlorococcus* cells as their hosts. A virus is simply a rudimentary set of genes packaged in a protein coat. In order to reproduce, it has to inject its DNA into a host cell, take over its machinery, and use it to make more virus particles. Sometimes the viral DNA loops into the host genome and assumes a holding pattern until the time is right to take over the host and make more virus particles. Viral and host genes get exchanged on occasion, and if not detrimental to host or virus, the shuffled gene persists through generations in the recipient. It may even become an asset over time. What is an asset to a virus, of course, has to be a liability for the host cell it successfully infects. But that cell has, we assume, many identical clones in the ocean—sister and daughter cells—that will carry on its heritage and likely never see that particular virus again. So one can think of a host cell as simply playing a role in maintaining the diversity of the federation by keeping the gene shufflers in business. And gene shuffling, over time and space, might just be a key contributor to the global diversity, structure, and stability of the *Prochlorococcus* federation.

When I look at our emerald green laboratory cultures of captive *Prochlorococcus*, my mind quickly turns to their wild cousins, drifting freely in the world oceans. I am reminded that while they carry out a respectable fraction of the photosynthesis on our planet, they escaped our attention until a few decades ago. What else of this magnitude are we not seeing? Will we find it before human activities have completely dominated the oceans, as they have the land? As I write this, commercial ventures are gearing up to fertilize the oceans to trigger phytoplankton blooms, designed to draw carbon dioxide out of the atmosphere to abate global warming. If carried out on grand scales this approach would, by design, dramatically alter the marine food web. I have repeatedly voiced the position that commercialized ocean fertilization is an ill-advised climate mitigation strategy. Some have argued, only partially in jest, that my protests are simply a disguised concern for *Prochlorococcus*. On the contrary, these tiny cells occupied our planet long before humans, and they will surely outlast us. They can photosynthesize. They thrive through diversity. Their federation can adapt. And, as one of my graduate students once put it, they have a time-tested strategy: "Grow slowly, and endure."

Because this essay was written for the nonscientist, I omitted specific attributions for ease of narrative. In using the editorial "we" I am assuming the role of spokesperson for the many scientists who were involved in this story. There are a few people who played pivotal roles in the initial chapter of "the unveiling," however, who should be recognized: Robert Olson, using a flow cytometer, was the first to notice the "little greens"; Ralph Goericke and John Waterbury played central roles in putting the puzzle together; and Brian Palenik isolated them into culture. These contributions set the stage for all that followed. Since those early days, there have been many—too many to mention by name—who played key roles in giving Prochlorococcus *a voice. May they continue to do so!*

Microbes and Evolution: The World That Darwin Never Saw
Edited by R. Kolter and S. Maloy

doi:10.1128/9781555818470.ch24

24

Deciphering the Language of Diplomacy
Give and Take in the Study of the Squid-*Vibrio* Symbiosis

Margaret McFall-Ngai and Ned Ruby

The Geological Context of the Evolution of Animal-Microbial Partnerships

The colonization of epithelia by microbes is the most common and ancient form of animal symbiosis. All of the different major groups of animals, from the sponges to those with backbones, had their origins in the world's oceans. When they came on the scene, about a billion years ago, the oceanic environments were already organically rich: they had been colonized by microbes for over 2 billion years, so they were full of the microbes themselves, as well as the

Margaret McFall-Ngai obtained her Ph.D. at the University of California at Los Angeles (UCLA), and then did postdoctoral research at Jules Stein Eye Institute (UCLA) and Scripps Institution of Oceanography. She has had tenured faculty positions at the University of Southern California, the University of Hawaii, and the University of Wisconsin-Madison, to which she moved in 2004. Her idea of a fine time is to spend hours in a modern art museum. She also has a passion for tennis, body surfing, and skiing. Her hero is Dorothy Parker. *Ned Ruby* did postdoctoral training at Harvard, Woods Hole Oceanographic Institute, and UCLA after completing his thesis at Scripps Institution of Oceanography. His faculty positions have been at the University of Southern California, the University of Hawaii, and (since 2007) the University of Wisconsin-Madison. He enjoys basketball and looks forward to body surfing whenever he can get to the ocean.

by-products of living and dead microbial cells. The architecture of the different types of animals (i.e., their body plans) evolved to take advantage of this situation. Most notably, animal body surfaces, which are composed of epithelial tissues, evolved to function in two ways: (i) to take up dissolved organic material and use it in their nutrition, and (ii) to manage relations with the microbes sharing their habitat. In the former function, marine animals competed with the microbes for the dissolved nutrient pool, while in the latter function, animals could either encourage or discourage direct interactions of their epithelia with the environmental microbes. Many animals combined these two functions in symbiosis. They formed either transient or persistent alliances wherein metabolically clever microbes would help them access organic and inorganic nutrients (i.e., C, N, and P) more efficiently. Throughout the evolutionary history of animals, from these early beginnings through the invasion of the land, and into the current biosphere, the most common type of animal-microbe association has been the colonization of the outer surfaces of animal epithelia by populations of one or more microbial species. Such associations are most often established anew each generation, a process that begins when the juvenile host leaves the protected embryonic environment (e.g., the egg or womb) and enters the microbe-dominated world.

Aided by new technologies, biologists have learned in the last decade that these beneficial alliances with microbes are critical for the biology of many animals. However, symbioses can be difficult to study experimentally, as they are often complex interactions of many microbial species with a single host. One strategy that biologists use to study such complex phenomena is to look for relatively simple systems and, through the study of these models, unravel the basics behind a process. Ideally, the system is one that is amenable to experimentation.

To decipher the language between a host and its symbiont during the colonization process, you'd want to be able to raise the partners independently in the laboratory, so that you could manipulate their interactions. The goal of such studies would be to find answers to such questions as: (i) how do the host and symbiont manage to get together in the first place—is timing important, and is the number of symbionts in the environment important? (ii) How do they recognize one another; i.e., how do they know their specific partner and interact with that partner to the exclusion of all others? (iii) How do they change in their form and function in response to the interaction? (iv) How do they achieve a balance so that the symbiosis stays healthy for both partners? For the last 20+ years, biologists have been studying the processes by which bacteria colonize animal epithelial surfaces, using the model association between the Hawaiian bobtail squid, *Euprymna scolopes*, and the marine bioluminescent bacterium *Vibrio fischeri* (Fig. 1).

The study of this association evolved as "symbioses" between biologists from two very different fields, microbiology and animal biology. How did these collaborations begin, and what have we learned about beneficial host-microbe associations from studying the remarkable squid-vibrio partnership?

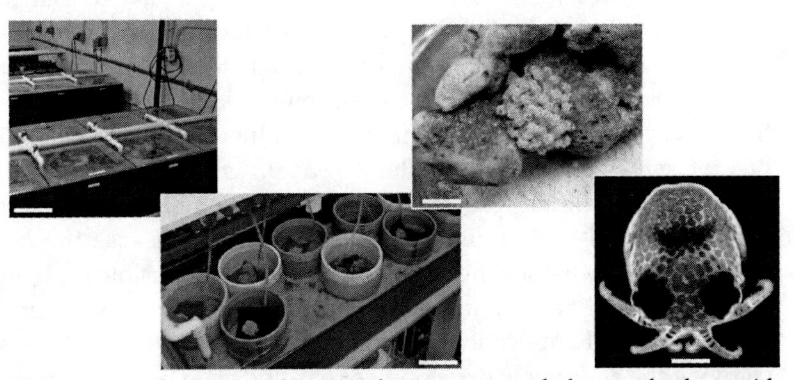

Figure 1 Generating the animal resources needed to study the squid-vibrio symbiosis. Male and female *E. scolopes* adults are maintained in individual seawater "condominiums" (upper left, the "condo" complex), where they mate and lay eggs (bar, 0.25 m). The females lay the eggs on hard substrates in their environment and cover them with sand. In our culturing facility, within 1 day of the eggs being laid, they are removed into a separate tank (lower left), so that they are not exposed during their embryonic period to symbiotic *V. fischeri* cells released from the adults (bar, 0.1 m). An individual clutch (upper right) of about 200 eggs has been deposited by a female on a piece of coral rubble (bar, 1 cm). A juvenile *E. scolopes* (lower right) will hatch from each egg after an embryonic period of ~20 days. In this photograph, the two lower darkened areas are the eyes and the upper dark area is the nascent light organ, which can be seen through the translucent body surface (bar, 0.5 mm). doi:10.1128/9781555818470.ch24f1

A "Symbiosis" Begins: Margaret's Story

My dissertation at the University of California, Los Angeles, in comparative physiology focused on the biology of ponyfishes, a group of fishes that inhabit the western Pacific and Indian oceans. I worked in the laboratory of James Morin, and my study site was the central Philippines, where I worked during the spring quarters of my graduate career. In their light-emitting organs, the ponyfishes have a symbiotic association with populations of the bioluminescent bacterium *Photobacterium leiognathi*. The combination of fascinating biology, the quality education guided by a rigorous doctoral committee, and the cultural experiences in Southeast Asia made my graduate career highly rewarding. However, a key research element was missing: the ability to manipulate the partners of this symbiosis to study the development of the symbiotic tissues. Just as I was thinking that I might have to change the research direction of my career, another door opened. While at a meeting, I learned from an oceanographer at University of Hawaii, Dick Young, about *E. scolopes*. Some evidence was available that this system was also a symbiosis with luminous bacteria, and that the symbiont was likely to be *V. fischeri*. As a bacterial partner, this species would be a bit of good luck—at the molecular level, it was among the best-understood marine bacteria because the *lux* operon, the cassette of genes encoding proteins responsible for light production, had been characterized and exploited in research and biotechnological applications. Further, Dick told me that biologists studying the embryonic development of squids had used *E. scolopes* as a subject. This feature of the host confirmed that it could be raised under laboratory conditions.

Although everything pointed to the squid-vibrio system as offering a great opportunity, one major problem presented itself. To be able to study this symbiosis experimentally, I would have to do something that I'd not done before, i.e., manipulate the bacterial symbiont. I quickly realized that the very best science would be done not by trying to do it all myself but, rather, by forming a "symbiosis" with a bacteriologist. Ned Ruby agreed to consider this adventure with me, and a symbiotic collaboration was under way.

Within the first year of working on the system, we realized that it is an ideal subject for the study of symbiosis. It would tell us not only about luminous bacterial associations but also about the most common type of animal-bacterial association: the colonization of epithelial cell surfaces. We found that the symbiosis has a series of biological features that render it an unusually tractable experimental model. A very small colony, 15 to 20 adult female and male animals, will, each month, generate ~5,000 juveniles that can be used for experiments (Fig. 1). In contrast with most other host-microbe associations, which can require weeks to years to come to maturation, the development of the squid-vibrio symbiosis naturally occurs over a few hours to days. This feature allows us to define hour by hour the interaction between the partners. In addition, the onset and progression of the symbiosis can be followed noninvasively by monitoring light production by the animal. Further, the juvenile animals are small (1 to 2 mm), so we can watch the symbiosis form either under a dissection microscope or by confocal microscopy (Fig. 1). Because of these key characters of the system, we have been able to use it to address each of the questions mentioned above.

Evolution of a Microbiologist: Ned's Story

Many of us were first drawn to biology by the remarkable diversity of animal life. Distinct species have evolved strange new abilities that allow them to inhabit even the most forbidding of environments. For me, the evolution of bioluminescent marine animals seemed a particularly rich field of study, combining ecology, behavior, and biochemistry. However, because I had never taken a course in microbiology, it was a surprise to learn that, unlike terrestrial light-emitting animals, many of the bioluminescent species in the sea had evolved symbioses with luminous marine bacteria to serve as their source of light. I began my doctoral studies with Ken Nealson at Scripps Institution of Oceanography, University of California, San Diego, who had a long-time interest in bacterial bioluminescence and was beginning to study the luminous bacteria present in the light organ of the reef-dwelling Japanese pinecone fish. It was from him that I learned about the fantastic physiological diversity that existed within the microbial world, and as a result, my frame of reference turned from animals to bacteria. As our work expanded to include luminous bacteria isolated from the light organs of many other species of fish, we began using a bacterial taxonomy approach newly developed by John Reichelt and Paul Baumann at the University of California, Davis. By applying their methods we were able to show that all the associations appeared to be species specific: that is, each fish species always had the same species of luminous bacterium in its light organ. However, we did not know whether this specificity resulted from a passage of the symbionts from parent to offspring through the egg or, instead, whether the light organ of a newly hatched juvenile could only be inoculated by one of the several species of luminous bacteria present in the surrounding seawater.

It wasn't long before I realized that for me, the most fascinating adaptation in marine luminous fishes was not the bioluminescence itself but the ability of an animal and a bacterium to function as a unit, resulting in a "hybrid" life form that created a biologically important product that neither could by itself: a persistent source of visible levels of light. I felt our best hope of learning how these symbioses evolved was by understanding how the two partners interacted; however, the light organ association of a mature host was too complex and intertwined. If only we could watch how the association developed, as was so elegantly being done in the nitrogen-fixing rhizobium-legume symbioses, we might be able to dissect the process step by step. Unfortunately, because no marine luminous fish produces offspring in the lab, it would be impossible to study how the bacterium and its host initiated their association. For this reason, when my doctorate was completed, I left the world of bioluminescent symbioses and worked on understanding the physiology of a series of other fascinating bacterium-specific adaptations like chemoautotrophy and periplasmic growth. But my desire to understand how symbiotic host-microbe interactions evolve remained.

It was over 10 years before Margaret rekindled that desire when she told me that the bobtail squid would lay eggs in the lab and that when juvenile bobtail squid hatch they have no symbionts, but their light organ can be inoculated by luminous bacteria in the seawater. With the support of these seminal studies, as well as the help of Dick Young and Dave Karl, a graduate school classmate of mine who was also at the University of Hawaii, Margaret and I collected our first bobtail squid and identified the symbionts as *V. fischeri*. Working out of the Hawaii Institute of Marine Biology and later the University of Southern California, where we were both on the faculty, we and our students began examining the initial stages of this symbiosis, discovering a remarkable bacterium-induced development and host specificity.

What We've Learned, Where We Are Now, and Where We're Going

Critical to the development of this research over the past 20 years has been the willingness of our students and collaborators to join us in this adventure. This effort has required the development of a variety of tools for the system. A major advance in capability came through a collaboration with researchers working on epithelial pathogens at the University of Iowa, Pete Greenberg, Mike Apicella, and Mike Welsh. With funding from the W. M. Keck Foundation, we developed genomic tools that have been used to describe the differences and similarities between how pathogenic and beneficial bacteria colonize host mucous membranes. For the squid-vibrio association, this effort yielded the first genome sequence for the bacterial partner and a clone library representing all of the genes expressed by the animal during the symbiosis. These important tools have allowed us to begin deciphering the molecular language of symbiosis between bacteria and animal epithelia. It has also revealed some very important clues as to how animal symbioses might coevolve. Because there are several host species that harbor *V. fischeri*, and many of those hosts can be cultivated, future genetic studies of *V. fischeri* promise to reveal how symbiotic partners have coevolved into the set of hybrid organisms that we see today, and that will provide an updated analogy to Darwin's initial concept of hybridization.

If we had to pinpoint the most significant contribution of this symbiosis thus far, we would have to say that "there's nothing new under the sun." We find that the molecular language of this mutualistic symbiosis is, in many ways, very similar to that of pathogenic symbioses. That is, the strategies and molecules used are often the same, and it is the context that makes the difference in what the outcome will be. These findings, coupled with the recognition that mutualistic symbioses are far more prevalent than pathogenic ones, invite biologists to go back to first principles to rethink their basic ideas concerning the mechanisms underlying, and consequences of, animal-bacterial hybridization.

FURTHER READING

McFall-Ngai M. 2008. Host-microbe symbiosis: the squid-vibrio association—a naturally occurring, experimental model of animal/bacterial partnerships. *Adv Exp Med Biol* **635:**102–112.

McFall-Ngai M. 2008. Are biologists in "future shock"? Symbiosis integrates biology across domains. *Nat Rev Microbiol* **6:**789–792.

Ruby EG. 2008. Symbiotic conversations are revealed under genetic interrogation. *Nat Rev Microbiol* **6:**752–762.

Visick KL, Ruby EG. 2006. *Vibrio fischeri* and its host: it takes two to tango. *Curr Opin Microbiol* **9:**632–638.

Microbes and Evolution: The World That Darwin Never Saw
Edited by R. Kolter and S. Maloy

doi:10.1128/9781555818470.ch25

25

The Tangled Banks of Ants and Microbes

Cameron R. Currie

> It is interesting to contemplate a tangled bank, clothed with many plants of many kinds, with birds singing on the bushes, with various insects flitting about and with worms crawling through the damp earth, and to reflect that these elaborately constructed forms, so different from each other, and dependent on each other in so complex a manner, have all been produced by laws acting around us.
>
> Charles Darwin, *On the Origin of Species* (1859)

This quote, commonly referred to as the "tangled bank," is one of the most well-known passages in Darwin's *On the Origin of Species*. When reading it, either to myself or when giving a lecture, I am reminded of the complexity of our natural world. For me, this passage brings to life the concept that organisms do not occur in isolation; instead, they are embedded in complex communities, composed of a diversity of forms that are interconnected through webs of interactions. In part, Darwin's genius is his recognition that over time, the "constructed forms" of species are largely shaped by their interactions within the ecological communities in which they occur. Over the 150 years since the publication of Darwin's seminal

Cameron Currie received his Ph.D. from the University of Toronto. He was first hired as a faculty member in the Department of Ecology and Evolutionary Biology at the University of Kansas before moving to the Department of Bacteriology at the University of Wisconsin-Madison. He enjoys traveling to tropical countries to collect ants, and being a proud Canadian, he loves hockey.

treatise, countless studies have demonstrated that interactions among species are largely shaped by, and are a driver of, evolution by natural selection. Simply put, it is difficult to overstate the importance of species interactions in biology; they influence all levels of biological organization, from genes to ecosystems.

One particularly important interaction is the formation of intimate associations between two or more unrelated organisms, a relationship referred to as symbiosis. I should note that although some biologists limit the use of the term symbiosis to beneficial associations (mutualism), I use the term in the broader sense (including parasitism, where one partner benefits at a cost to the other, and commensalism, where one partner benefits at no cost to the other). My passion as a scientist is in studying the complexity, origins, and evolution of multipartite symbiotic interactions that occur within the tangled banks of ecological communities. Here, through the exploration of my main study system, an ant-microbe symbiosis, I highlight the complexity of ecological communities associated with individual organisms—what I refer to as tangled banks embedded within tangled banks.

The tangled bank I am most fascinated by—some might say obsessed with—is not composed of birds singing in the bushes or worms crawling through the earth but, instead, is entrenched within the colonies of a specific group of ants: the Attini. Attine ants occur only in the New World and are commonly referred to as fungus growers. As indicated by their common name, they are distinguished from other ants by their ability to cultivate fungus in specialized gardens. The queen, her brood, and most of the workers live on and within the fungus garden, which is typically maintained in subterranean chambers. Workers carefully tend the fungus and forage for substrates to use as nutrients to support its growth. In exchange, the fungus serves as the primary food source for the colony. Functionally, this is similar to human agriculture, but instead of growing plants, the ants grow fungus. This is an obligate mutualism, as the ants cannot survive without the fungus and vice versa. The interdependence of the ants and their fungus is linked across generations; to found new colonies, an incipient queen carries a small pellet of fungus from her parent nest, which she uses as an inoculum to start her own fungus garden.

This ant-fungus mutualism is ancient; we believe it originated in the Amazon basin of South America ~45 million to 55 million years ago. Following the formation of the initial ant-fungus partnership, the ensuing millions of years of this association are characterized by diversification of both partners. The evolution of the ant-fungus mutualism culminates in the captivating and conspicuous leaf-cutter ants (Fig. 1).

Leaf-cutters collect fresh plant material, mostly leaves, to cultivate their fungal mutualist. The foraging activity of these ants is so prodigious that they are one of the most dominant herbivores of Neotropical ecosystems. Individual colonies of some species can live for more than 10 years, be composed of millions of workers, and harvest more than 400 kg (dry weight) of fresh leaves a year. After over a decade of studying these ants, I still find it awe-inspiring to watch a worker cut a leaf disc, hoist it over her head with her mandibles, and then join a long trail of thousands of workers carrying leaf fragments back to the colony.

Biologists have been studying fungus-growing ants for more than a century. My own adventures with these amazing insects

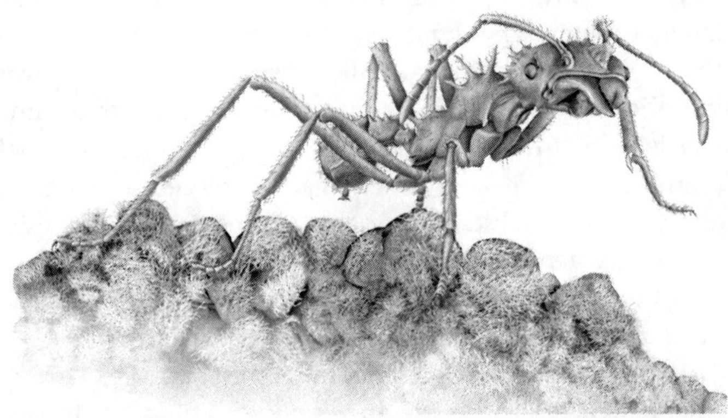

Figure 1 A leaf-cutter ant worker standing on the surface of a fungus garden. Leaf-cutters have a mutualistic association with fungi they cultivate for food. Workers forage for fresh leaf material, which they use to support the growth of the fungus. The fungus can be seen protruding as threadlike projections off the leaf substrate of the garden matrix. This line drawing was created by Angie Fox. doi:10.1128/9781555818470.ch25f1

began in the latter half of the 1990s. I read, in a number of the classic papers on the system, that the ants were so adept at agriculture that they maintained their gardens free of parasites. In exploring the literature more deeply, I realized that evidence to support this was largely lacking and that the mechanisms employed to defend the garden from exploitation were poorly resolved. This intrigued me, so I decided that for my doctoral thesis I would look for the presence of specialized parasites and explore the mechanisms of garden defense.

What I found, perhaps not too surprisingly, is that contrary to the popular dogma of the time, the fungus gardens are in fact susceptible to infections by parasites. Fungi in the genus *Escovopsis* are special pathogens of the ant-fungus mutualism and can use the tissue of the ants' cultivated fungus as its primary nutrient source (Fig. 2).

Although *Escovopsis* sometimes completely overgrows entire fungus gardens, thereby killing the colony, typically infections are chronic. The persistent presence and continuous growth of *Escovopsis* within the fungus garden matrix cause a significant decrease in the growth rate of the garden and indirectly reduce the number of workers produced by the colony. Having discovered a virulent parasite of the fungus garden, I then focused on mechanisms the ants might employ to protect their fungus and discovered several previously unknown methods of garden defense. First, worker ants use their mouthparts to carefully clean the fungus garden, a behavior we now refer to as "fungus grooming." Second, similar to the way in which humans deal with weeds, worker ants actively excise and discard infected fungus garden pieces, a process called "fungus weeding." Third, these ants engage in a symbiosis with bacteria in the genus *Pseudonocardia,* which can completely cover the bodies of worker ants and produce antibiotics that help protect their fungus gardens from *Escovopsis* (Fig. 2).

Pseudonocardia are members of the Actinobacteria, a group well known for their ability to produce potent antibiotics. Indeed, the majority of antibiotics used in human medicine are derived from Actinobacteria. So, it is perhaps not surprising that through natural selection fungus-growing ants "discovered" the benefit of obtaining antibiotics from these bacteria. This relationship, however, is

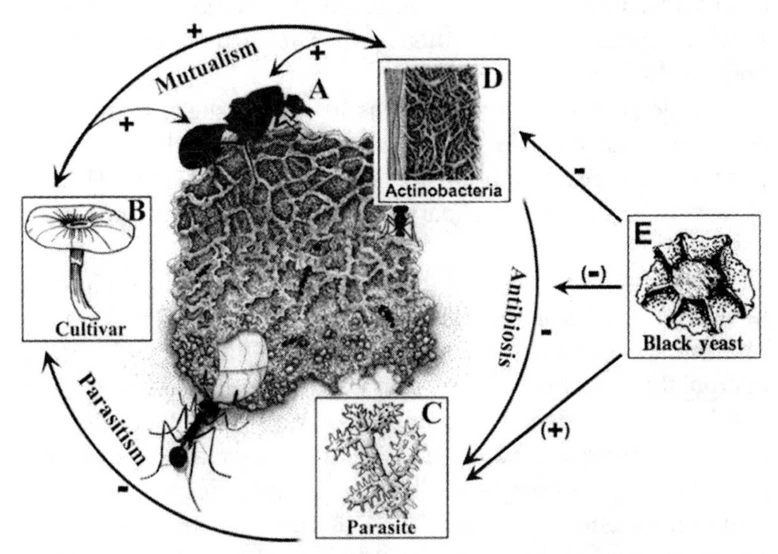

Figure 2 The tangled bank of ants and microbes within the fungus-growing ant-microbe symbiosis. (A) A leaf-cutter ant queen, shown sitting on the top of a fungus garden, can live more than a decade and lay ~60 million eggs over her life span. Individual leaf-cutter colonies can be composed of millions of workers and hundreds of fungus garden chambers. (B) The fungus serves as the primary food source for the ant colony and is related to soil-decomposing fungi, which produce mushroom fruiting bodies similar to what is depicted in the box. (C) The fungus garden is often infected by specialized parasites (genus *Escovopsis*), which continuously threaten the success of the ant-fungus mutualism. (D) The ants have a mutually beneficial association with bacteria (*Pseudonocardia*) that produce antibiotics that suppress the growth of the parasite. (E) A fifth symbiont, black yeast, exploits the ant-bacteria mutualism, consuming the bacteria and decreasing its effectiveness at suppressing the parasite. (Arrows represent the interacting components, with plus and minus signs indicating whether the interaction is beneficial or antagonistic. For antagonistic interactions, the head of the arrow points to the organism that experiences the negative impact.) The original line drawings were created by Cara Gibson and Rebeccah Steffensen. doi:10.1128/9781555818470.ch25f2

not one-sided, as *Pseudonocardia* benefits from its symbiosis with the ants. Specifically, the bacteria gain access to the specialized niche of the ant colony, an environment relatively free of competitors. In addition, like the fungus, the bacteria are dispersed to new colonies by founding queens. Furthermore, some ant species have evolved specialized and often elaborate structures (crypts) for housing *Pseudonocardia* (Fig. 3).

These crypts are invaginations in the exoskeleton of the ants and are connected to specialized gland cells that apparently provide nutrients to support the growth of the bacteria. Thus, this system is, minimally, a tripartite mutualism between ants, fungi, and bacteria.

As with the ants and their fungi, *Escovopsis* and the ant-associated *Pseudonocardia* are composed of species-rich lineages. Multiple lines of evidence indicate that the diversity of these two microbial symbionts is the product of millions of years of evolution within the ant system. For example, our studies on the evolutionary history of *Escovopsis* have revealed that the most recent common ancestor of *Escovopsis* was likely a parasite of the most recent common ancestor of the fungi the ants cultivate. This indicates that *Escovopsis* was likely present at the very origin of the ant-fungus mutualism and suggests that when the ants first domesticated their fungal mutualist, they also inadvertently acquired the parasite. Moreover, the broad evolutionary history of *Escovopsis* is highly congruent with that of the ants and their fungal cultivar, a pattern of codiversification that supports an ancient association of *Escovopsis* with the system. Finally, *Pseudonocardia*'s ancient and specialized association with attine ants is supported by the presence of the crypts on workers, which vary depending on the ant species in both their location and form. In summary, our studies indicate that over their long evolutionary history, attine ants have been continuously threatened with famine induced by *Escovopsis* infections of their fungus garden; to help mitigate this threat, the ants have been employing antibiotic-producing bacteria for millions of years.

Recently, we have discovered the presence of a fifth symbiont lineage in the system: black yeast, which exploits the ant-*Pseudonocardia* mutualism. The black yeast occurs on the bodies of worker ants in the same location as *Pseudonocardia* and is capable of using the bacteria as its sole nutrient source. These findings indicate that the yeasts are parasites of *Pseudonocardia,* although in addition to consuming the bacteria, they may also usurp nutrients provided to the bacteria by the ants. We have been unable to detect any direct

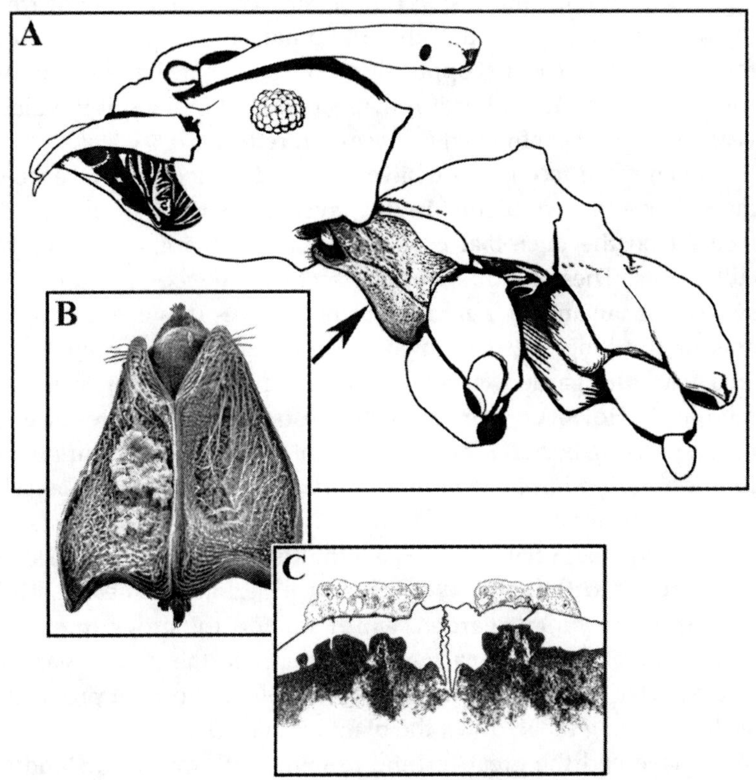

Figure 3 Line drawings of the symbiosis between fungus-growing ants and the antibiotic-producing bacteria (*Pseudonocardia*). (A) Side view of the head and thorax of a worker ant. Just below the head are the propleural plates (arrow), the main location of bacterial growth in most ant species. (B) A close-up of the propleural plates, illustrating the growth of the bacteria (left plate) and the specialized crypts evolved to house the bacteria (right plate, after removal of bacteria). (C) Cross section through the propleural plates, illustrating the large biomass of bacteria within the crypt (bottom), the cuticle of the ant (middle), and specialized gland cells that apparently provide nutrients to support the growth of the bacteria (top). Line drawings created by Sarah Taliaferro. doi:10.1128/9781555818470.ch25f3

impact of the black yeast on the health of the ants. However, the yeasts do compromise the ants' ability to deal with the garden parasite, apparently by reducing the abundance of *Pseudonocardia* and, concomitantly, the production of antibiotics (Fig. 2). Interestingly, our findings that *Escovopsis* obtains an indirect benefit from the black yeast suggest that these two parasites might themselves form a mutualism in opposition to the tripartite ant-fungus-bacterium association, but it remains to be determined if the black yeast obtains a benefit from *Escovopsis* in return.

Even more recently, we have started detailing the presence of more bacterial symbionts in the system, work that is largely focused on the microbial ecology of the fungus gardens of leaf-cutter ants. These fungus gardens serve as the external digestive system for the ants; the garden converts leaves into energy for the ants. It has long been assumed that nutrient provisioning to the ants from the garden is only mediated through the ants' fungal mutualist. However, our work has established the presence of nitrogen-fixing bacteria that supplement the nitrogen requirements for the ants. We have shown that the bacteria in the genus *Klebsiella* are important nitrogen fixers in the fungus garden, and stable-isotope studies established that the fixed nitrogen indeed is integrated into the ants. We have since performed metagenomic studies on the fungus garden, which support the presence of four or five more genera of bacterial symbionts in the fungus garden. These bacteria appear to play additional roles in nutrient provision, such as helping break down the plant biomass.

I have had the great fortune to work with many outstanding and talented scientists who have conducted studies on fungus growers that are only partly motivated by our fascination with these ants. More importantly, we hope to help inform on our general understanding of evolution, the unifying theory of biology. Here, I describe a few ways in which our work relates to the field of evolutionary biology. First, studies on symbiotic associations, such as ours, are critical for gaining insights into coevolution. Coevolution is a process where two organisms exert selective pressure on each other (i.e., they affect each other's evolution). Indeed, in Darwin's tangled bank, the "laws acting around us" he refers to are coevolutionary forces that shape the constructed forms we see. The attine ant-microbe symbiosis illustrates the potential complexity of coevolution with not just two partners coevolving but a whole community. Teasing apart the evolution of this community will likely provide additional insights into coevolution. Second, this system illustrates the role of symbiosis as rapid evolution. Through the formation of symbiotic associations with their fungus and bacteria, the ants obtain access to the metabolic capacity of these microbes, including the capacity to degrade plant biomass and produce antibiotics, respectively. This is what some refer to as "symbiosis as evolutionary innovation," which I believe is a far more important evolutionary process than is currently recognized. Third, the discovery of a mutualism between the ants and *Pseudonocardia* suggests that Actinobacteria may have evolved to play similar roles with other insects. Indeed, we recently discovered a species of Actinobacteria that helps mediate the fungal community associated with a species of bark beetle, analogous to the role *Pseudonocardia* plays in the attine symbiosis. Finally, the long-term stability of mutually beneficial associations, such as the ant-fungus mutualism, is an evolutionary paradox. As Darwin himself recognized, selection should favor individuals that pursue their own selfish interests, exploiting their partner by obtaining benefits without providing a reward in return. The discovery of a parasite exploiting the ant-fungus mutualism raises the possibility that it helps stabilize the mutualism by aligning the interest of the ants and their fungi. Indeed, experimental work we have conducted has shown that if either the ants or fungi cheat on their cooperative partner, the garden parasite has a much more significant impact on the mutualism. This suggests that the community of interaction associated with cooperative organisms may help stabilize their association.

As described above, embedded within the colonies of fungus-growing ants is a diverse community of microbial symbionts. These symbionts span bacteria and fungi that include beneficial and antagonistic associations with both direct and indirect effects. This illustrates the potential complexity of microbial communities associated with a single insect. This is just one example of the many recent studies that have revealed that individual macroscopic organisms harbor communities of symbiotic microbes. When you next read or hear Darwin's tangled-bank passage, perhaps you will visualize, as I do, the diverse communities of microbial symbionts embedded within each individual insect, plant, worm, or bird. I cannot help but think that Darwin would be fascinated by recent insights microbiologists have made into the "unseen" tangled banks embedded within the tangled bank he contemplated over 150 years ago.

Microbes and Evolution: The World That Darwin Never Saw
Edited by R. Kolter and S. Maloy

doi:10.1128/9781555818470.ch26

26

Microbial Symbiosis and Evolution

Nancy A. Moran

The first time I read Charles Darwin, as an undergraduate in Texas, I wondered why so much of his famous book was about pigeons, and domesticated species and breeds in general. By chance, I ended up studying pigeons myself in my undergraduate research project, specifically, how females picked mates. But my interests soon moved from the sexual habits of pigeons to asexual habits of all-female species of insects and then to anciently enslaved bacteria, all in the course of studying evolution and how it works.

What Darwin didn't realize is that our oldest domesticates are within our own cells. Virtually every cell of our bodies contains organelles, called mitochondria, all of which descend from a bacterial cell that, 2 billion years ago, was engulfed by another cell. During my undergraduate years, this monumental fact regarding the history of life was still considered an unproven hypothesis. Now, thanks to the molecular biology revolution, it is no longer under dispute.

Darwin recognized that traces of our evolutionary past are surprisingly near to hand, reflected in our own anatomy and even our facial expressions. What he couldn't know was that even the

Nancy Moran studied at the University of Texas and the University of Michigan. She was a professor at the University of Arizona from 1986 until 2010, when she moved to Yale. She had to give up raising chickens when she moved, but now is happily harvesting honey from hives kept (mostly) for research on bee symbionts.

very earliest stages of evolution are embedded within our own cells and genes. Of course, he was limited by a major gap in understanding, since the physical nature of heredity was not yet known. It would be another 100 years or so until the roles of nucleic acids, DNA and RNA, would be discovered. But once they were, and once efficient technology was developed for reading the history embedded in these giant molecules, nucleic acids presented us with a trace of the whole panorama of evolution, from the origins of life over 3 billion years ago to the paternity analyses in humans, birds, and insects.

We know that symbiosis, the very intimate cohabitation of a microbe and a host, was pivotal both in the earliest stages of evolution and, later, in the twigs and branches of the vast tree of life. The big event was a single acquisition of an oxygen-respiring bacterium by another cell; this bacterium evolved to become the modern-day cell organelle called the mitochondrion. The mitochondrion is a little sausage-shaped structure present in almost all cells of people, plants, and many other species; it is the energy generator for our cells, allowing us to make good use of oxygen for our various energetic pursuits. In the case of mitochondria, we can each thank our mothers, who are generally the exclusive source for these essential items.

The second big event was the uptake of yet another bacterium by one of those mitochondrion-containing cells, the ancestor of algae, green plants, and seaweeds. This new symbiont was a photosynthetic bacterium able to harvest carbon dioxide and sunlight to make energy and molecular oxygen. Plants spread and filled the air with oxygen, and this allowed organisms with mitochondria to breathe deep, expend energy, and grow bigger, giving rise to animals, some, like ourselves, having big brains requiring efficient energy supply. Plants also produced the giant carbon deposits that we are now burning in order to drive to the local gym.

Thus, symbiosis was critical for the evolution of animals and plants in particular, that is, for almost all of what most people would recognize as living things. DNA studies provided the proof and also details of who engulfed whom and how often. In fact, it now appears that there was only a single original infection each, for mitochondria and for chloroplasts. Before these two critical events, Earth's atmosphere had little molecular oxygen and no large creatures of any kind. Life existed but was sluggish. (Of course, even slugs were yet to be.)

So what has become of our oldest domesticates, the mitochondria? These hard workers are only barely recognizable as bacteria. They have lost most of their parts, including most of their genes. When first ingested and kept by our unicellular ancestor, about 2 billion years back, they had perhaps 2,000 to 3,000 genes. Our mitochondria today have only 15. You might wonder how we know that they are bacteria at all. The few genes that they do contain, by virtue of their DNA sequences, point very clearly to their origin from a particular group of bacteria, a group that today contains soil bacteria, pathogenic bacteria, and many marine bacteria. So why did they lose almost all their genes? First, they don't need most of them because we, their hosts, provide for most needs. Mutations, including deletions of DNA, happen constantly, so the unused genes have long since disappeared from the genomes of mitochondria that inhabit our cells. In fact, mutations are still raining down on the genomes of our mitochondria, and these continue throughout our lives, contributing to various diseases and disorders. Indeed, these malfunctions are the main reason we even think about our mitochondria.

Another reason the mitochondria can live with so few genes is that many of the ancient genes of the mitochondrial ancestor have changed addresses, moving from the chromosome of the mitochondrion to our own chromosomes but sending their working products back to the mitochondrial homeland to carry out needed work. For example, one essential job is the copying of the DNA that makes up the little circular chromosome with its 15 genes. This task is carried out by an enzyme that is encoded in the host genome and produced in the host cell, but transferred to the mitochondrion.

The mitochondrial story has several lessons. One is that the host, or master, evolves to take care of the symbiont, or slave, and that both end up entirely dependent on one another. The reasons are obvious: hosts with defective symbionts will die or fail to reproduce and thus will be removed by natural selection from their population. A second lesson is that genes not needed are genes lost.

We now have many more examples of bacteria that became symbiotic in a bigger organism and then lost most of their genes. It turns out that cases of symbiotic bacteria that are passed from mother to babies are quite common. Somehow a bacterium enters the host, lives happily within her, and then manages to infect her eggs or babies before they are born. Extend this lifestyle over generations, and the bacterium has a stable life. No need to locate new patches with nutrients and proper conditions: instead, the symbiont stays in the host and all is provided. Of course, the host needs to stay alive and reproduce; otherwise, the bug encounters a dead end. So bacteria that are inherited from mother to offspring are strongly favored to avoid harming their hosts and even to actively benefit the host. For their part, bacteria have a lot of tricks, due to their abilities to make many compounds and to break down many others, so they have many possible skills that can benefit a host. For example, they can make toxins that kill enemies of the host. They can make nutrients that the host needs but cannot make itself. And they can break down materials, like cellulose, into food usable by the host.

Aphids are insects that feed on plant sap. They have the interesting property of being able to reproduce without the benefit of males or sperm. Inside a single female aphid, about half a million bacterial cells reside. The great majority of these are one particular type, called *Buchnera*. These are rather large bacteria with tiny genomes, only about 600 genes altogether. *Buchnera* lives inside some special cells of the aphids; any *Buchnera* organisms that escape these cells are broken to bits. This system gives the aphid control over its *Buchnera* population. *Buchnera* is required for aphids to grow and reproduce. Antibiotics kill *Buchnera* and ultimately the aphids as well. What does *Buchnera* contribute? Plant sap is an attractive diet in some respects, because the plant brings the sugar-rich food to the aphids, which sit in one place and simply suck. But this food is deficient in a number of nutrients required by all animals, from aphids to humans, namely, the 10 essential amino acids. *Buchnera*'s role is to make these amino acids. In fact, *Buchnera* is a little amino acid production machine, having lost about 85% of its ancestral gene set, including the genes that underlie most of its abilities to make other new compounds, but keeping exactly the set of genes needed to make the nutrients the hosts require. Mutations in *Buchnera* continue to happen. If one of these lands in a gene needed for making a particular amino acid, the aphid host can be affected, either dying or suddenly requiring that amino acid in its diet.

Compared to mitochondria, the *Buchnera*-aphid symbiosis is recent. Still, *Buchnera*'s ancestor entered an insect host about 200 million years ago, still ancient by most time scales. And these millions of years have been sufficient time for *Buchnera* to undergo massive change and to evolve an intimate and codependent relationship with its host.

In the case of the needed amino acids, they are produced in only one way, using a set of genes that evolved very early in Earth's history. For example, there is only a single way to make tryptophan, via enzymes encoded by seven specific genes. Since all organisms require tryptophan, either they have these seven genes or they get their tryptophan from another organism that has them, either by eating that other organism (the human solution) or by absorbing tryptophan secreted by a symbiont (the aphid solution). To do its job, *Buchnera* merely has to maintain, in good working order, its ancestral genes for making tryptophan and other amino acids.

But what about genes not needed by *Buchnera* or its host? Darwin wrote at length about the atrophy of organs or structures that are no longer needed. Now that we can study evolving genomes, we can see atrophy at the gene level. How does a gene, no longer needed, change and decay? The basic answer is that mutations happen, and they are generally destructive. Just as a car that sat in the driveway for decades would gradually lose its ability to start up and go, genes unused can only degenerate over time. Of course, mutation-based decay threatens any genome and any gene. But this decay is greater in symbionts because they have such specialized habits that many genes are just not needed, or not needed very much.

In fact, symbiotic bacteria have the tiniest genomes of any organisms (viruses don't count as organisms, in my view). Some symbiotic bacteria have fewer than 200 genes, and the symbionts we call organelles can have even fewer, as in our own mitochondria with their miserly 15. Much of the reason is that the symbiont lives within a host and leaves the host to do most of the job of producing cellular parts.

Like symbionts, domesticated animals depend on hosts (us) for most of their needs and can thereby afford to lose some of their ancestral prowess. A white Leghorn chicken is probably not as smart and certainly not as able to survive in the wild as its ancestor the jungle fowl. But it does lay more eggs. Similarly, *Buchnera* cells explode within minutes of removal from their host aphids, but they have special adaptations for making extra nutrients for their hosts. For example, *Buchnera* uses its ancestral genes for making the amino acids tryptophan and leucine, which are nutrients that the aphid host needs. But these genes have been copied many times on tiny extra chromosomes (plasmids), an arrangement that allows hyperproduction of the amino acid products. This resembles milk production by a dairy cow: she uses the same machinery as did her wild ancestors; she just invests more and makes more milk. Both symbionts and domesticated breeds can evolve extreme features over short periods on an evolutionary time scale. These extremes are produced mostly by exaggerating some functions, rather than by acquiring anything truly novel. And just like our domesticated animals and plants, symbionts themselves can be large, typically much larger than their free-living wild relatives. This is because they are so protected by their host that they can do without the usual powers of transport and sturdiness.

These examples show that symbionts, passed from mother to babies, have different roles and limitations. They may feed their hosts or breathe for them. Some protect hosts from parasites and pathogens or from heat stress. But symbionts also can be the basis of their hosts' demise, as they mutate and thereby lose genes and functionality. In the latter case, they provide yet another reason to blame mom, the usual source of symbionts, good or bad.

Microbes and Evolution: The World That Darwin Never Saw
Edited by R. Kolter and S. Maloy

doi:10.1128/9781555818470.ch27

27

Coevolution of *Helicobacter pylori* and Humans

Martin J. Blaser

In 1983, Robin Warren and Barry Marshall published the isolation for the first time of a gram-negative bacterium that lives in the human stomach, and the next year they suggested an association with peptic ulcer disease. We now call that microbe *Helicobacter pylori*, and for their isolation of *H. pylori* and their studies linking its presence to peptic ulcer disease, Warren and Marshall received the 2005 Nobel Prize in Medicine. One of the many remarkable features of *H. pylori* is that after being acquired during childhood, it usually persists for life in its hosts.

Much of the early work, including our own studies beginning in 1985, was focused on whether and how *H. pylori* could induce inflammation in the stomach. By the early 1990s, it was becoming clear that carriage of *H. pylori* increased risk for peptic ulcer disease and for gastric cancer, which both are consequences of the *H. pylori*-induced gastric inflammation, although in separate ways. Treatment to eradicate *H. pylori* changed the natural history of peptic ulcer disease. Yet in most hosts, there was no disease, only persisting colonization. How could this organism that was so proinflammatory, especially in vitro, persist for decades in the

Martin Blaser studied economics before training in Medicine at NYU and Colorado. He is a faculty member at NYU in the departments of Medicine and Microbiology. He loves nature and thinking about how nature solves problems. Hiking, travel, art, culture, geography, and history help with perspective.

stomach in virtually all hosts? This is the conundrum that I have been considering for many years. In this chapter, I provide some of the scientific aspects of this question that my colleagues and I have explored.

I would like to begin this story by going back to a very rainy day in March 1990, when I was attending a University of Leeds meeting about *H. pylori* in Harrogate, a 19th-century resort in northern England. There in my lovely and charming hotel room, while the rain cascaded from the heavens, I thought that the ability of the organism to persist cannot just involve its upregulation of inflammation; I hypothesized that there must be downregulation as well. In a Newtonian sense, these must be balanced, in some form of equilibrium. On a sheet of paper, I sketched out this idea. Unlinked signals between host and microbe select for an evolutionary arms race, but if the signals are linked, with negative feedback, then equilibrium can be achieved. This is a model of homeostasis similar to those present in endocrine circuits; my paper that included this concept was published in February 1992.

Over the next few years, I worked with Denise Kirschner, a very talented mathematician, to formalize the concept and to determine whether equilibrium was achievable under the parameters related to *H. pylori* that were known at the time. Our initial model had four populations. The first two were *H. pylori* organisms that adhered to the gastric epithelium and those that were free-living in the gastric mucus layer. We believed that it was necessary to distinguish between these because the adherent and free-living cells likely would signal the host differently and would be differentially affected by host signals, such as peristalsis. The other two populations were those of the effector molecules that *H. pylori* sends to the host and the host release of nutrients induced by the *H. pylori* effectors that could sustain the bacterial population. The model worked; using known ranges of parameter values, we could achieve long-term equilibrium.

In the next iteration, we incorporated a fifth and essential population which is based on the host's immune responses. This term reflects the host innate or adaptive responses that limit the microbial burden. With this model, we observed an initial transient state with increased bacterial loads that diminished with the onset of active immunologic memory, to a lower level that then could persist in steady state. We could show the critical effects of immunity, as well as the dynamics of competition between individual strains. Our work identified the importance of privileged niches that would allow individual strains to persist without being outcompeted by a strain that is generally more fit. Later studies focused on the role of strain-specific restriction endonucleases in protecting cocolonizing organisms from subversion by their *H. pylori* competitors, since the cells are naturally competent for DNA uptake.

We presumed that the equilibrium had been established long ago by the ancestors of present-day *H. pylori* and *Homo sapiens*. We began to examine that issue by studying *vacA*, a gene encoding a protein that can induce vacuoles in epithelial cells. Although strains vary in their expression of the VacA protein, the gene is conserved in all strains; we found that the gene has two alternative signal peptides that we called s1 and s2, with s1 strains producing stronger in vitro activity. In studies of the geographic distribution of s1 *vacA*, we found that East Asian strains were often of a type that we called s1c, whereas strains from Europe were mostly s1a or s1b. We showed a North-South cline, with s1a highly prevalent in northern Europe and s1b most common in southern Europe, such as the Iberian Peninsula. These studies indicated that there is a geographic structure of the present *H. pylori* population.

Based on these findings, we reasoned that when we examined Latin American isolates, we would mostly find s1c, reflecting the origins of Amerindians in Asia and the crossing of the Bering Strait. In large coastal cities, we found that the strains were s1a and s1b, which was surprising. However, then we realized that the sampled populations were mestizo (of mixed European, African, and Amerindian ancestry) and that if we wanted to sample Amerindians, we should study people living in the interior of South America. We sampled *H. pylori* strains from Amerindians deep in the Orinoco-Amazon basin, by working with Maria Gloria Dominguez Bello, a Venezuelan biologist who has long studied this group. There, we found strains with the s1c allele, as well as other markers of Asian origin. This finding indicated that as early humans migrated across the Bering Sea land (ice) bridge, which last was extant about 11,000 years ago, they brought *H. pylori* in their stomachs to the Americas, and that the descendants of these strains have been maintained in present-day Amerindians.

In recent years, using more sophisticated analytical techniques, there has come greater support for the notion that *H. pylori* has colonized humans since before the out-of-Africa migrations of about 58,000 years ago, and that as humans migrated to all parts of the world, they brought their *H. pylori* strains with them. Such stability of *H. pylori* and its universality in human populations support the notion of coevolution between microbe and host.

However, a model of coevolution based on equilibrium implies that rules have been set and that all parties follow these. Yet natural selection predicts that "cheaters" will emerge who will break the rules for their own benefit and with these advantages then topple the system. The emergence of cheaters is a major constraint on thinking about coevolution. And yet examples of coevolution abound in nature in which two, or even more, species have learned to cooperate to maximize their own fitness. How could this have been accomplished, and in particular, how did *H. pylori* and humans evolve to their relationship of persistence?

Part of the solution is to understand that extinctions occur. Nature has no investment in failed experiments, and those individuals or communities that cannot compete are lost. In that sense, cooperation arose in part because the alternatives became extinct, and the cooperative systems were robust to withstand change. John Maynard Smith developed the concept of evolutionarily stable strategies to describe cooperating systems that can resist invasions, ideas that have been extended by other scientists.

More recently, Denise Kirschner and I also have explored explanations for the equilibrium between *H. pylori* and humans. The pioneering work of John Nash in game theory appeared most applicable. He described circumstances, now known as the Nash equilibrium, under which it is disadvantageous to defect; the cheater is actually in a worse position than individuals who play by the rules. We hypothesized that the coevolutionary relationships characterizing several bacteria with persistent interactions with humans (*H. pylori, Salmonella enterica* serovar Typhi, and *Mycobacterium tuberculosis*) conform to Nash equilibria. However, Nash equilibria only exist under particular circumstances, and the boundary conditions are critical. We hypothesized that for *H. pylori* and humans, there are actually a series of equilibria, nested in one another, that create the governing boundaries at each level. It is a solution that requires multiple biological scales to link the fates of the two competing organisms: microbe and host.

From this construct, we can hypothesize that the equilibria with *H. pylori* were established during the long period (~2 million years) of human existence in small groups as hunter-gatherers. Person-to-person transmission was intensive, and if the ancestor of present-day *H. pylori* aided early-life (through reproductive age) survival of its human hosts, there would have been strong selection of humans for its maintenance. In that sense, the vertical, and to some degree horizontal, transmission of *H. pylori* would facilitate host survival. Postulated benefits for humans for early-life carriage of *H. pylori* include resistance to colonization by exogenous pathogens (through manipulation of gastric pH and gastric immunity, as well as direct competition), as well as close regulation of metabolism, through gastric leptin (5 to 10% of body total) and ghrelin (60 to 80% of body total).

Finally, microbes that kill their hosts late in life (after the end of the reproductive period) may actually be selected since their host population groups have fewer mouths to feed. The log-linear development of *H. pylori*-induced gastric cancer is a form of biological clock that can safely (for the population as a whole) remove senescent individuals. If *H. pylori* indeed has early-life health benefits and late-in-life health costs, it may be viewed as a highly selected symbiont.

Yet, despite its long coexistence with humans, *H. pylori* has been going extinct over the course of the 20th and now 21st centuries. Currently, fewer than 10% of native-born children in developed countries are acquiring and maintaining *H. pylori*. This is an astonishing change in human microbiology. At a time when epidemic diarrheal disease and famine no longer threaten children in developed countries, the benefits of having *H. pylori* in our children's stomachs may be reduced. Whether or not humans are better off with or without *H. pylori* is not yet resolved, and the answer to this question may vary according to host and locale.

In summary, *H. pylori* has evolved over a long period of time as a highly interactive member of the human gastrointestinal microbiota. There is extensive evidence that *H. pylori* coevolved with its human hosts, enabling its nearly universal gastric persistence. The interaction had little or no cost (and possible some benefit) to its early-in-life hosts, but also conferred certain late-in-life disease costs, including gastric cancer. However, despite its long-term biological successes, *H. pylori* is becoming extinct due to the pressures of modern human life, a remarkable change in human microecology. Not surprisingly, the *H. pylori*-free stomach appears to have substantially different physiology than the colonized stomach. In future years, we will better learn the consequences of this major shift in our "genetic" heritage.

This work was supported in part by R01GM63270 and UH2 AR057506 from the National Institutes of Health and by the Diane Belfer Program for Human Microbial Ecology. I thank the many coauthors and colleagues whose work I have mentioned, and the many others in this active field of research who have contributed so much to our knowledge.

FURTHER READING

Blaser MJ, Atherton J. 2004. *Helicobacter pylori* persistence: biology and disease. *J Clin Investig* **113:**321–333.

Blaser MJ, Kirschner D. 2007. The equilibria that allow bacterial persistence in human hosts. *Nature* **449:**843–849.

Franks SA, Schmid-Hempel P. 2008. Mechanisms of pathogenesis and the evolution of parasite virulence. *J Evol Biol* **21:**396–404.

Lenski RE, May RM. 1994. The evolution of virulence in parasites and pathogens: reconciliation between two competing hypotheses. *J Theor Biol* **169:**253–265.

Moodley Y, Linz B, Yamaoka Y, Windsor HM, Breurec S, Wu JY, Maady A, Bernhoft S, Thiberge JM, Phuanukoonnon S, Jobb G, Siba P, Graham DY, Marshall BJ, Achtman M. 2009. The peopling of the Pacific from a bacterial perspective. *Science* **323:**527–530.

Microbes and Evolution: The World That Darwin Never Saw
Edited by R. Kolter and S. Maloy

doi:10.1128/9781555818470.ch28

28

The Library of Maynard-Smith
My Search for Meaning in the Protein Universe

Frances H. Arnold

I'll never forget reading Jorge Luis Borges' short story "The Library of Babel" when I was working in Madrid in the summer of 1976. I was thrilled by Borges' description of this collection of all possible books and deeply struck by the despair of the librarian, who searched for meaning in the essentially infinite stacks of "senseless cacophony, verbal nonsense, and incoherency." I found it delicious that the librarian's certainty that one of those books contained the answers to all the fundamental mysteries of humankind was accompanied by the unbearable realization that he or she could never locate such a treasure. He had to accept finding a few phrases embedded in typographical gibberish as the sum of his life's work. I could not imagine spending life wandering that desolate maze.

Two decades later, Dan Dennett's wonderful book *Darwin's Dangerous Idea: Evolution and the Meanings of Life* brought back these youthful memories and put them into context for me in a powerful way. Dennett's variation on Borges' library, the Library of Mendel, is the collection of all possible genomes (whose written description, by the way, is a subset of the Library of Babel). And what glorious

Frances Arnold has been a chemical engineer (Ph.D. from University of California at Berkeley) at the California Institute of Technology for many years. She enjoys thinking about protein evolution and microbiology when she is not working on alternative energy or hiking to her rustic cabin in the San Gabriel mountains.

creatures Mendel's library contains! But the vast majority of possible genomes does not encode life—as has been noted, there are far many more ways to be dead than alive.

I have spent my years searching for meaning in one of the virtual rooms of the Library of Mendel, the Maynard-Smith Collection, which houses the set of all possible protein sequences. Like the Libraries of Babel and Mendel, the Maynard-Smith Collection is incomprehensibly large, and the vast majority of its sequences encode rubbish. Instead of 500 pages, a typical protein is only 500 letters long, and there are only 20 letters (amino acids) in the protein alphabet (for now, at least). But that gives 20^{500} different combinations, already well beyond the number of particles in the universe. Meaningful proteins—proteins that *do* something—are extremely rare, because there are far more ways to be meaningless (e.g., unfolded and unable to function). I can revel in all the gorgeous proteins that must be contained within this collection—cures for cancer, the answer to the energy crisis. But have I joined the Babel librarian in a hopeless search for these magnificent but extremely rare books?

Maynard-Smith's library is different from Borges' in two very important ways. These differences have made it a far more satisfying universe to explore. The Babel librarian's deep depression came from his or her inability to find even a comprehensible sentence, much less a book, even a bad one, written in any recognizable language. He was also frustrated by the apparent random order of the books. Reading one book would give no clue as to the contents of the next, and therefore no directions as to where one should go in order to find a better book. In the libraries of Mendel and Maynard-Smith, meaningful books jump right out at us. They are alive! You can literally scrape some of these rare genomes, ones that actually encode life and all its glorious proteins, from the bottom of your shoe! (Of course there are many other interesting and meaningful genomes and proteins, that you cannot find on your shoe, or anywhere on Earth for that matter, but at least we are not throwing ourselves off the balconies of the Babel Library for never having had the pleasure of reading such a book.) This is the great gift of natural selection, which has done the search for meaning for us. Natural selection relentlessly washes out the weak stories and supports the strong. And woe to those who do not heed natural selection's editorial dicta! They might become recipients of the infamous Darwin Awards, conferred—posthumously in most cases—on those who "do a service to humanity by removing themselves from the gene pool."

The other key difference between the Maynard-Smith (virtual) Library and the hopelessly unorganized Babel warehouse is that there is order in the former. Because he was not limited to three paltry physical dimensions, Maynard-Smith could organize his protein books in a very special way. Each sequence is surrounded by its one-mutation neighbors, that is, by all the proteins that differ from it by a change in a single amino acid letter. Never mind that there may be thousands of ways to change a single letter in a protein book—remember, this is a virtual library, not a physical one. Maynard-Smith set up his library with as many dimensions as there are ways to make a mutation.

Why is this ordering so important? These neighbors are the ones that evolution explores. When a random mutation—a step to a neighboring book—is made, natural selection "reads" that book. And, with natural selection inexorably sweeping away the debris, only the meaningful ones are left behind. A step to a bad book just sets you back to where you were before. Over millions of years, this is just what evolution has done, taken meandering paths through this vast space of possible proteins. Look at all the interesting molecules that have emerged! And we only see the ones that still exist, on Earth, in 2012. So many beautiful proteins have been discovered and incorporated into the living world. So many more have been tested and forgotten, or found wanting and thrown away. And many, many more have never been tested at all, for there has not been enough time, since the beginning of life on Earth, for evolution to have tested even a small fraction of this vast space of possibilities.

Now, imagine that I get to decide which books in this library are meaningful. I'm on a search for new proteins, proteins unlikely to be found in the natural world: proteins that might be useful to someone or that will tell me something unconventional that natural proteins, the few that have survived editing by natural selection, cannot. I'm the pitiless editor now, and I get to judge each book by my own criteria. I read each one, and if I don't like it, I throw it away and go back to my previous book. But if I do like it, I move on, taking another step in a random direction. With a single change, a mutation, the book may look a lot like the one before, but then again, it may not if that word is critical. (Dennett's example from *Moby-Dick*: consider the difference between "Call me Ishmael" and "Call me, Ishmael.") If I have a particular story in mind, I can even choose to move to books that begin to build that story. I can accumulate my choice of one-letter changes, and if I am patient I might even be able to follow this path to a pretty good new story. It may not be the "best" book in the collection, but remember that this is a very big library. I can't hope to find the best. But what I find may be good enough to keep me happy, especially if the alternative is the book I started with, or no book at all.

My students and I have spent the better part of 20 years taking sometimes straight and sometimes slightly crooked paths in Maynard-Smith's library of proteins. Maybe "forced marches" is a better way of describing our walks: these experiments can be tiring. It's a lot of work to read all these sequences—make all these proteins—and decide which are interesting and which are not. So my forays have been short. And few.

While I have explored but the tiniest fraction of this remarkable place, I can tell you some interesting things I have learned during my travels. First, even more than books do, protein properties can change quite significantly with even a single mutation. Proteins can become more stable, they can change their biological functions, or they can change color. Most of the possible mutations don't have a big effect, as far as I can tell with my limited eyes. And, of course, some of the mutations can greatly damage or even destroy a protein (and remove it from the gene pool). But we can often find mutations that allow the protein to do new things, such as catalyze a reaction on a new substrate or adapt to environments that it had not previously tolerated. I was gratified, perhaps even surprised, to see how "evolvable" these lovely molecules are, how quickly they could "learn" new tricks! Of course, now I understand that this property of adaptability is precisely what allowed protein-based life to take over and diversify into the many forms we enjoy today.

I also learned that proteins can do lots of things that nature probably never asked them to do, including things that humans might benefit from. A reviewer of one of my earliest proposals (back in the previous millennium) criticized my idea to direct the evolution of proteins to do unnatural things, like function in an organic solvent. He said that proteins simply couldn't do that, because nature had never gone there before. Of course, this is exactly the wrong reasoning. It's precisely because nature did not ask for these behaviors that we were so successful in finding mutations that allowed them. In fact, it's much harder to find mutations that will improve a protein doing its natural job, because nature has been looking for those for a while. But improvements in something that the protein does not naturally do well are usually far easier to find. There's much more room for improvement, and there are many more paths that lead to improved sequences, provided you don't ask for too much too fast.

I also learned that there are plenty of places that a hike in Maynard-Smith's library isn't likely to take you. Since I don't have the lifetimes to devote to my searches that Borges' librarians did, I have to be careful about which steps I take. There are lots of steps that take me away from the protein I am looking for and only a very few that appear to take me closer. I say "appear" because I cannot see beyond the protein in front of me; it's as if I am climbing a mountain in the clouds and can feel or see only one step ahead. I choose, therefore, to take the step that moves me closer to the top, but I can never know if this particular route will ultimately lead me there. So it takes some faith as well as patience.

This blind walk has nonetheless been highly productive. Since we understand so very little about how the sequence of a protein encodes meaning, that is, how its sequence determines what a protein does, we cannot write our own books. That, of course, is the dream of many a protein engineer, but I fear that it is a distant if not impossible one. We have to discover interesting new proteins by making them and seeing what they do. I have found that a walk in Maynard-Smith's collection is a wonderful way to discover these new proteins. There are exciting stories just waiting to be discovered.

FURTHER READING

Borges JL. 1941. *The Library of Babel.* Editorial Sur, Argentina.
Dennett DC. 1996. *Darwin's Dangerous Idea: Evolution and the Meanings of Life.* Simon and Schuster, New York, NY.
Smith JM. 1970. Natural selection and the concept of a protein space. *Nature* **225:**563–564.

Microbes and Evolution: The World That Darwin Never Saw
Edited by R. Kolter and S. Maloy

doi:10.1128/9781555818470.ch29

29

In Pursuit of Billion-Year-Old Rosetta Stones

Dianne K. Newman

If I have learned anything about evolution, it is that the path life takes moves in mysterious ways. If someone had told me 20 years ago that I would wind up a professor of geobiology, I would have laughed. It certainly wasn't something I aspired to in college, where, as a German studies major, I didn't take a single biology class and took only one geology class. I didn't even know that geobiology existed as a discipline at the time!

My undergraduate passion was an esoteric problem in the humanities. Having as a sophomore seen an art exhibit entitled "Entartete Kunst," which recreated a 1937 exhibit put on by Nazis contrasting the "degenerate art" of the Weimar Republic with the art that they promoted—a strange form of neoclassicism on steroids—I became fascinated by how Nazis had appropriated classical Greek culture for political purposes in the early 20th century. Accordingly, much of my undergraduate education centered on classes in art, classics, and German literature.

Because I also loved science, I filled the remainder of my units with classes in materials science, chemistry, and environmental

Dianne Newman did her academic training at Stanford, MIT, Princeton, and Harvard Medical School. Since 2000 she has been a faculty member at Caltech. Recently, she has been forced to revise her opinions about the relative importance of nature versus nurture by observing her young son's passion for "getting the bad guys" and turning everything possible into a sword.

engineering. I spent a good part of my sophomore year doing research on compounds used in magnetic recording devices, and as a senior, I joined a group that applied organizational theory to construction management. How this was preparing me for a career studying the evolution of microbial metabolism certainly wasn't obvious to me then.

As Darwin has taught us, evolution happens through natural selection. Small things, which in any given instance might not appear significant, can grow to have great importance when the environment changes. Realizing that I couldn't do much with a German studies degree, and eager for a change of pace, I enrolled in graduate school in environmental engineering because I thought it was practical; my intention was to get a master's degree and work for a few years as an engineer before possibly applying to law school. In my first semester, I took a course in environmental microbiology. I was amazed to learn about the incredible metabolic properties of microorganisms, which could "eat" things as toxic as toluene and "breathe" things other than oxygen. Not only did this strike me as inherently cool, but also it impressed me by its potential utility. Bacteria could be used to clean up the messes humans made of their environment if the conditions were right for them to grow! Not bad.

So when it came time for me to pick a research project the next semester, I petitioned my advisor (who was a chemist) to let me do something with bacteria. Being a generous spirit, he agreed, and he introduced me to a senior graduate student in the department who was studying bacteria that could breathe arsenic. Breathe arsenic? Whoa. How could that be? I had no idea, but I happily accepted a small airtight bottle from this student containing sediment from a lake north of Boston in which bacteria were growing and producing something yellow. She suggested I study chemotaxis towards arsenic, but not knowing what chemotaxis meant, I instead decided to figure out what the yellow stuff was, as I knew how to approach this thanks to my undergraduate research in materials science.

The yellow stuff turned out to be the semiconductor arsenic trisulfide (As_2S_3), also known as the mineral orpiment, from the Latin *auripigmentum,* meaning "golden pigment." Previously, this compound had been thought to be produced naturally in geothermal vents through strictly chemical reactions. Serendipitously, I had discovered that the bacteria in my bottle were producing it as an end product of their metabolism. And thus my career as a geobiologist began. My intellectual journey from arsenic trisulfide to the topics my research group studies today was enabled by the support of many people. Suffice it to say that I had a lot of catching up to do in the life sciences, and I was fortunate to meet people who encouraged my budding interests in microbiology, molecular biology, and geology and gave me room to grow. Now, many years later, I cannot imagine being anything other than a microbiologist, and I use every opportunity I get to "spread the good word" about bacteria and their environmental and evolutionary importance.

Microorganisms are amazing. They invented the metabolic machinery that sustains all life today. Indeed, the cellular compartments where energy is made in the human body and plants are nothing more than the remnants of ancient bacteria that were engulfed by and entered into symbioses with other cells long ago. Microbes are far more sophisticated metabolically than we are—they run circles around us. Whereas we can eat organic carbon and breathe oxygen, bacteria can eat and breathe just about anything, provided what they consume satisfies some minimum energy requirement. Not only are they the consummate omnivores, but also they thrive in every conceivable niche on Earth, from the ice of the Antarctic to the hydrothermal vents in the oceans' abyss. Pressure, pH, salinity, temperature, toxins...you name it, they have figured out a way to handle it through cunning biophysical and biochemical inventions.

Through geologic time, some of these inventions have been important on a global scale. For example, the oxygen in the atmosphere we breathe results from ancient bacteria figuring out how to convert water to oxygen during photosynthesis (the same process that plants perform today). Massive sedimentary deposits of iron and manganese formed either directly or indirectly from microbial activities. Today, microorganisms can be used for a variety of biotechnological purposes, from converting cellulose to ethanol to leaching precious metals from ore deposits. Their metabolism can be harnessed to fertilize plants with fixed nitrogen, generate energy to power light bulbs in Third World villages, consume nasty organics in contaminated aquifers, and precipitate toxic metals from solution before they reach potable water supplies. And these examples just scratch the surface of what bacteria can accomplish. Admittedly, not everything bacteria do is good for the environment. In some parts of the world, such as Bangladesh, their respiratory activities contribute to the release of arsenic into the groundwater that has led to a public health crisis of epic proportions. And of course, a few bad species are responsible for some major infectious diseases. But the vast majority of bacteria play important roles in sustaining both human health and the health of our planet.

So when and how did this remarkable metabolic diversity arise? As with most things, timing is everything. Practice makes perfect, and microbes have had a *lot* of time in which to develop and hone their skills. When most people think about linking events in evolutionary history to particular moments in time, they think of fossils: dinosaur bones, trilobites, a footprint captured in an ancient sand—the type of structures we can see on display in most natural history museums. But what I want everyone to understand is that morphological fossils, as impressive as they are, record but a *fraction* of the history of life. Indeed, while the familiar fossil record begins around 580 million years ago, the history of life arguably dates to at least 3.8 billion years ago (Ga). Who occupied all the time in between? Microorganisms, of course. Therefore, if one wants to understand biological evolution in deep time, one must embrace the microbes and the traces they leave behind. And therein lies the challenge: knowing what to look for that definitively can be interpreted as a biosignature, versus something generated by a nonliving process, is not trivial.

Until recently, a small band of intrepid geologists trekking over the remotest portions of the globe have led the way in identifying and interpreting signs of ancient life. For example, the argument that microbial life originated on Earth at least 3.8 Ga is based on ratios of different forms of carbon (i.e., isotopes) that are preserved in rocks of that age from western Greenland. Basically, the carbon in these rocks is enriched in a light isotope, something one would expect if that carbon originated from a living cell. This finding is controversial, as whether these rocks are really this old, whether they came from an environment that would have been hospitable to life, and whether the light carbon might instead have been generated through an abiotic process are debatable.

These caveats aside, several other lines of evidence—both geochemical and morphological—point to the early evolution of microorganisms. These include kilometer-scale ancient carbonate platforms, deposited some ~3.4 Ga, which are thought to be remnants of reefs built by microorganisms; ~3.4-Ga organic laminae that have been interpreted as fossils of ancient microbial mats; structurally complex organic molecules that are molecular fossils of certain types of lipids, found in rocks as old as ~2.7 Ga; and isotopes of sulfur that are recorded in ~2.4-Ga minerals that preserve a strong signature of microbes "breathing" sulfate in the ancient oceans. Taken together, these data provide a convincing record of the presence of microorganisms on Earth very early in its history.

What was the nature of these ancient life forms? What types of metabolisms kept them alive, and when and how did they evolve new capabilities? These are difficult questions to which we will never have definitive answers, short of the invention of a time machine. Nevertheless, they inspire geobiologists like myself to engage in speculation constrained by facts. For example, some of the students and postdocs in my laboratory work with genetically tractable modern bacteria possessing attributes that are likely to be ancient. Our hope is that by learning how these organisms manifest these attributes, we will be better able to interpret the rock record. For example, we are currently studying diverse photosynthetic bacteria because we are interested in when different types of photosynthesis evolved. These bacteria all produce the particular lipids that are the progenitors of the 2.7-Ga molecular fossils I mentioned.

Conventionally, when these molecules are identified from any given sample, they are interpreted as biomarkers of cyanobacteria—the microorganisms that invented *oxygenic* photosynthesis, the conversion of water to oxygen in sunlight—and hence, they are assumed to be biomarkers of oxygenic photosynthesis itself. But are they? The reason we care is that the invention of oxygenic photosynthesis not only was a biochemical tour de force but also changed the world, affecting the composition of the land, air, and sea. So it would be nice to know whether these molecules can accurately be used to date this invention or whether we are barking up the wrong tree. It increasingly seems that the lipid progenitors of these molecules have nothing to do with oxygenic photosynthesis, given that only a minority of modern cyanobacteria make them, and they are made by other bacteria that cannot engage in this metabolism. What else, then, might these molecular fossils mean? Even if they turn out to signify something very different, it's okay: once we crack the code, we will have learned how to decipher a class of molecules that tell us something important about evolution. The remaining challenge will be to identify new biomarkers to help constrain the timing of major metabolic breakthroughs, such as oxygenic photosynthesis, and shed light on how these breakthroughs came to be.

There is a book entitled *The Past Is a Foreign Country* by David Lowenthal, which I came across while writing my undergraduate thesis, that helps me rationalize my personal intellectual journey towards geobiology and summarize why I find it compelling. The message I take from it is that the past, whether it is captured in a 20th-century painting or in molecules from a 3.8-Ga-old rock, provides a mysterious allure and promise of discovery. The evolutionary events that shape history, whether human cultural history or the history of microbial metabolism, are riddles waiting to be solved. This requires creativity, rigor, tenacity, and—like evolution itself—random luck. In anticipation of the continued appearance of new clues, the development of increasingly powerful technologies, and the aid of talented colleagues, I have no doubt the pursuit of the molecular equivalents of billion-year-old Rosetta Stones will provide enough excitement to last my lifetime.

I thank the following people for enabling my career: Raymond Levitt, Katharina Mommsen, Gerald Gillespie, Lee Krumholz, Dianne Ahmann, Francois Morel, Terry Beveridge, Abigail Saylers, Ed Leadbetter, John Stolz, Ron Oremland, Bonnie Bassler, Tom Silhavy, Dale Kaiser, Roberto Kolter, Edward Stolper, and Jonas Peters. I am grateful to my colleagues, especially the members of my research group, who have made the discovery process so enjoyable.

FURTHER READING

Knoll AH. 2003. *Life on a Young Planet.* Princeton University Press, Princeton, NJ.

Lowenthal DL. 1985. *The Past Is a Foreign Country.* Cambridge University Press, Cambridge, United Kingdom.

Newman DK, Gralnick JA. 2005. What genetics offers geobiology. *Rev Mineral Geochem* **59:**9–26.

Schopf JW (ed). 1983. *Earth's Earliest Biosphere.* Princeton University Press, Princeton, NJ.

Sessions AL, Doughty DM, Welander PV, Summons RE, Newman DK. 2009. The continuing puzzle of the Great Oxidation Event. *Curr Biol* **19:**R567–R574.

Van Kranendonk MJ. 2006. Volcanic degassing, hydrothermal circulation and the flourishing of early life on Earth: a review of the evidence from c. 3490–3240 Ma rocks of the Pilbara Supergroup, Pilbara Craton, Western Australia. *Earth-Sci Rev* **74:**197.

Microbes and Evolution: The World That Darwin Never Saw
Edited by R. Kolter and S. Maloy

doi:10.1128/9781555818470.ch30

30

The Deep History of Life

Andrew H. Knoll

The view down Lake Louise is easily one of North America's finest, with towering peaks framing crystalline blue waters. Standing on the lakeshore, I sense the same majesty as other hikers, but I also see something else—something equally grand, but in a different way. Our planet records its own history, written in sedimentary rocks deposited one layer on another through geologic time. In the mountains behind Lake Louise, then, I see a library, and if we know how to read its volumes, we can glimpse our world in formation.

I was first attracted to sedimentary rocks as a boy growing up in the foothills of the Appalachian Mountains. On weekends and summer days I loved nothing more than to split open limestones and shales, pulled from local hills and road cuts, to reveal fossil shells or leaves inside. I can't tell you which was more exciting, the sight of organisms that lived millions of years ago or the act of discovery, of unearthing fragments of life's history that no one had ever observed before. As I grew up, my twin loves of fossils and discovery persisted undiminished, although my quarry diminished quite a bit, at least in size. From a childhood infatuation with dinosaurs, I graduated to a more realistic (in central Pennsylvania)

Andrew Knoll earned his Ph.D. from Harvard University. Following five years on the faculty at Oberlin College, he returned to Harvard, where he teaches biology and Earth science. He enjoys cooking and fieldwork, but not, it must be said, cooking in the field.

search for brachiopods and corals, older, if less scary, than *Tyrannosaurus rex*. Then, as a college student, I discovered a record of life still older—and much, much smaller.

The conventional fossil record is built of hard parts—bones, shells, and decay-resistant organic tissues buried in the sediments that accumulate on floodplains, in lakes, and on the seafloor. As he wrote *On the Origin of Species*, Charles Darwin was keenly aware of this record—and bothered by it. In particular, Darwin was unsettled by the oldest known fossils, trilobites preserved in mudstones near the base of the Cambrian System in Wales. Trilobites have bodies of impressive complexity—segmented bodies, jointed legs, and compound eyes—and so, Darwin reasoned, could not record Earth's earliest life. They must have been preceded by simpler forms that evolved complexity gradually over vast oceans of time. Where was the record of these earlier organisms? Darwin speculated that it lay buried beneath younger rocks, had been destroyed by erosion and metamorphism, or simply remained to be discovered in remote regions unvisited by Victorian geologists.

For nearly a century after publication of *On the Origin of Species* (1859), the oldest fossils remained animal skeletons in Cambrian rocks. A few brave paleontologists reported microscopic fossils in older beds, but such claims were routinely dismissed by a skeptical community. Take, for example, a 1939 critique by Harvard paleontologist Percy Raymond: "Walcott leaves it to be accepted on faith that an organism without hard parts, and less than 0.001 millimeter in diameter, would be preserved in identifiable condition from pre-Cambrian time to the present!" Raymond's choice of punctuation tells us all we need to know about contemporary views on Precambrian fossils.

In the 1950s, things began to change. First, geologists began the routine application of radioactive decay to problems of geologic age. The radioactive breakdown of unstable isotopes had been used to estimate the age of fossils as early as 1913, but only in the 1950s did laboratory innovations enter into productive partnership with geologic mapping. As a result, geologists came to understand that the oldest Cambrian fossils were about 542 million years old, but also that Earth itself formed more than 4.6 billion years ago. Continuing research has pushed the age of the oldest known animals to about 650 million years, but what came earlier? As geologists began to calibrate our planet's history, biologists discovered a molecular basis for building a tree of life that depicts the evolutionary relationships of all organisms, from elephants to *Escherichia coli,* redwoods to rotifers. Animals form only a small cluster of branches near the tip of one limb. By implication, then, the greater diversity of life, and life's deep evolutionary history, must be microbial.

The recognition that Earth's earliest inhabitants were microorganisms raises a particular challenge for paleontologists. Bacteria may be ancient and ubiquitous, but they are also tiny and seemingly fragile; could Earth's early microbial inhabitants have left a discernible signature in our planet's oldest sedimentary rocks? Remarkably (to most) and excitingly (to me), they could and did. In fact, the geologic record of microbial life is preserved in four distinct ways. First, bacteria and protists leave what we can consider an extension of the conventional fossil record: cell walls and extracellular envelopes preserved directly in sedimentary rocks. In 1953, my mentor, Elso Barghoorn, and his colleague Stanley Tyler first reported unambiguous microfossils of bacteria in Canadian rocks nearly 2 billion years old, nearly quadrupling the known history of life.

Second, microorganisms also leave molecular fossils that complement the record of morphology. Just as conventional fossils comprise decay-resistant remains like shells and bones, so, too, do some biological molecules have a good chance of avoiding decay. Unfortunately for paleontologists, proteins and nucleic acids rarely enter the geologic record; they are simply too good to eat. Lipids, however, are a different story. When you die, the last component of your body to disappear may well be the cholesterol laced through your cell membranes (and probably lining your arteries). Lipids made by bacteria, archaea, and unicellular eukaryotes occur abundantly in petroleum and other ancient organic deposits.

Sediments transported across the seafloor interact physically with microbial mat communities, providing a third and distinctly different biological signature in sedimentary rocks. Especially when the precipitation of mineral cements causes microbially influenced layers to accrete vertically into three-dimensional structures called stromatolites, the signature of ancient microorganisms can be massive. Microbial reefs in Precambrian sedimentary successions can be as large as those built today by corals and algae. Finally, microbial populations can actually influence the composition of seawater, providing a distinct chemical signature in minerals precipitated from ancient oceans. For example, when photosynthetic organisms incorporate carbon dioxide into organic matter, they preferentially use CO_2 containing ^{12}C, the lighter stable isotope of carbon. Because of this, limestones and organic matter deposited on the seafloor preserve a record of primary production in the sea above them. In similar fashion, bacterial respiration using sulfate instead of oxygen preferentially uses SO_4^{2-} containing ^{32}S, the lightest isotopic form of sulfur, imparting a chemical record of the biologic sulfur cycle into gypsum and fool's gold formed in the oceans and underlying sediments.

The records of ancient microbial life, then, may be more subtle than dinosaur bones strewn across a valley floor, but they are no less abundant if you know how to look for them. For much of the past three decades, I've worked to discover and read these paleobiological records of Earth's oldest life, and to do so in a systematic way that reveals evolutionary patterns. I've spent long summers climbing cliffs in the Arctic, piloting rafts down Siberian rivers, and swatting flies in the outback of Australia; the quest to understand life's early history remains an act of exploration, in both time and space. But the record is there, and it makes sense in light of predictions from the tree of life. Sedimentary rocks just older than those that contain the earliest known animals preserve protozoa and simple algae, as well as cyanobacteria and other prokaryotes that must have been ubiquitous in shallow seas. Double the antiquity, to more than 1.5 billion years, and records of prokaryotes remain abundant, while those of eukaryotic microorganisms grow sparse. Double the age again, to more than 3 billion years, and the record becomes fragmentary and hard to interpret with confidence. One fact, however, stands out: the oldest sedimentary rocks available for study contain a clear signature of life. Earth has been a biological planet since its youth—and for most of our planet's long history, that life was microbial.

We can hazard only broad guesses about the biological properties of early microorganisms, but we can make one key statement with confidence: early cells lived without oxygen. The evidence for this is ironclad, both figuratively and literally. Distinctive rocks called banded iron formations (BIF) occur in many sedimentary deposits formed before 2.4 billion years ago. These rocks, major sources of the iron used industrially, cannot form in modern oceans for the simple reason that iron is effectively insoluble in oxygen-rich seawater and so cannot be transported by ocean currents. Ancient iron formations show that early oceans were different; they must have been oxygen free throughout most of their depths, allowing iron to move in solution from place to place before precipitating out at sites of BIF deposition. Other geological observations corroborate this view, but about 2.4 billion years ago, the sedimentary record began to change. Rock chemistry tells us quite clearly that oxygen was beginning to accumulate in the atmosphere and surface oceans, and biology makes it clear that the source of that oxygen was cyanobacteria, the bacterial inventors of what biologists insist on calling "green plant" photosynthesis.

The world that emerged, however, was not our own but, rather, a long-lived intermediate biosphere in which moderately oxygen-rich surface oceans lay above subsurface waters that tended to be anoxic and, indeed, rich in sulfide. How do we know this? Iron, again, provides our principal clues to Earth's middle age, as iron minerals in seafloor muds record the chemistry of overlying waters. When, then, did our oxygen-rich world take shape? Only, it seems, about 600 million to 550 million years ago, just as animals with high rates of oxygen consumption began to appear in the fossil record.

In short, the story that has emerged piece by piece, from field research in the Arctic, in Australia, in Siberia, and in a number of other places, is one with three grand chapters. In chapter one, the first 2 billion years of our planet's history, oxygen was sparse but iron abundant; this is the world in which life emerged, and these were the environments that shaped the most fundamental features of biochemistry and ecology. Then, beginning about 2.4 billion years ago, the page turned to chapter two, Earth's long-lasting middle age, with moderate amounts of oxygen in the atmosphere and surface oceans but, commonly, sulfidic waters beneath that oxic veneer. Eukaryotic cells evolved, expanding the ecological possibilities of microbial communities, but bacteria continued to dominate the carbon cycle. Eukaryotic microorganisms began their rise to ecological prominence about 800 million years ago, as the sulfidic floor to the surface ocean began to dissipate, and with a second increase in oxygen levels 600 million to 550 million years ago, our familiar world of animals and plants began to take shape.

Decades after leaving the Appalachians, I still get a thrill from fossils and continue to enjoy the detective work of reconstructing evolutionary history from the biological and chemical details of ancient rocks. Conventional fossils—the bones and shells that fired my boyhood imagination—tell a good story, but it is only the latest installment of a much longer tale. Most of life's history, at least 85% of it, is a history of microorganisms. The plants and animals so conspicuous in our own world are evolutionary latecomers, intercalated into ecosystems that were already 3 billion years old when sponges first gained a foothold on the seafloor. While the pattern of life's deep history has become clearer, however, we still don't understand fully the processes that drove the transitions between long-lasting states of the Earth system. Life certainly played a role, but so did tectonic changes on our dynamic planet. I suspect that the correct explanation will not point to physical or biological processes acting alone but, rather, will emphasize the *interactions* between Earth and life. It has been, and continues to be, that interplay that guides evolution and environmental change through time.

Life, then, is not a feature of the Earth that stands apart from rocks, water, and air. Rather, life is an integral part of a highly interactive Earth system. Bacteria and archaea, with their dazzling diversity of metabolic processes, established ecosystems early in our planet's history, cycling carbon, sulfur, and nitrogen through primordial oceans. And more than 3 billion years later, despite millions of plant and animal species, microorganisms remain the fundamental cogs in the dynamic Earth system. Earth is a microbial planet—always has been and always will be. We, on the other hand, are evolutionary upstarts, recent and quite possibly transient arrivals whose continuing well-being will depend in no small part on understanding how our actions influence the microbial world we inherited.

Microbes and Evolution: The World That Darwin Never Saw
Edited by R. Kolter and S. Maloy

doi:10.1128/9781555818470.ch31

31

A Glimpse into Microevolution in Nature
Adaptation and Speciation of *Bacillus simplex* from "Evolution Canyon"

Johannes Sikorski

Why should one study microbial evolution? You might expect some sophisticated answer such as "it is intellectually fascinating" or "it is the queen of biological sciences," since "nothing in biology makes sense except in the light of evolution," to cite the famous quote of Theodosius Dobzahnsky. My motivation, as I retrospectively realize, was somehow different. As a Christian, I vaguely felt in my younger years that belief in God as the ultimate creator would not be commensurate with evolutionary theory as presented by science. Still, a talk by a German proponent of intelligent design and creationism couldn't fully convince me of that position. How could I know if the creationists' claims were indeed scientifically justified, when they contradicted the position of the overwhelming majority of scientists worldwide? So, besides my general interest in biology, I decided to study biology from scratch and to make up my own opinion. I admit that I expected to find enough evidence to prove on a rather professional level the evolutionary biologists to

Johannes Sikorski received his Ph.D. at the University of Oldenburg, Germany. After his postdoc in Haifa, Israel, he joined the DSMZ in 2006. He loves ambitious table tennis, running, and chasing his kids. For some relaxing moments, a good book, some wine, and classical music is just fine.

be wrong. To my surprise, I realized that the arguments of evolutionary biologists were not as wrong as I initially believed. By contrast, they made a lot of sense. At some point, I had to decide whether to accept the evolutionary theory as the (currently) best way to understand the "how" of biological diversity on a scientific basis, or to confess not to be amenable to reasonable arguments. You may understand that I decided on the former. Today, my personal understanding is that Christian belief is not defeated by the acceptance of the theory of evolution, nor is it necessary to give up Christian beliefs while doing evolutionary science. Interestingly, as a result of my efforts to understand scientific claims on evolution deeply enough to refute them, I got so attracted by this topic that I continued to work in this field.

So, let us go to the part of microbial evolution I am interested in. We do know that bacteria (and archaea) are genetically, phylogenetically, and physiologically very diverse. Despite the fascination with the macroevolutionary development of the very different bacterial phyla, my personal interest lies at the other end of the spectrum, at the birth of diversity. How do the first subtle and tender lineages accrue? Oh, you might say, that's trivial. Have a look at the textbooks and you will find everything about the evolutionary interplay of mutation, recombination, natural selection, and genetic drift. Much about these forces is addressed in this book by the other authors. The theoretical framework of population genetics is extremely well developed. Even more, you insist, microbial microevolution has been and is still being analyzed in very elegant laboratory experiments in which microbes are allowed to mutate, adapt, and evolve in test tubes under very stringent and therefore reproducible and adjustable conditions (see the contributions of, e.g., Richard Lenski and Paul Rainey). Right you are. But I might kindly remind you that the majority of bacteria are not evolving in a computer, nor are they living in the pencils of mathematicians and theoretical population geneticists or in laboratory test tubes, although all these tools have yielded tremendous insights. Most bacteria live outside in the environment, in water, soil, rocks, plants, etc. Here, where they face the enormous plethora of biotic and abiotic challenges, they evolve and speciate. Shouldn't we have a look at how evolution happens in nature, even though we as passive observers cannot steer and thereby control this process? Is there any chance to have a glimpse into such a natural evolutionary experiment? What are the decisive factors?

Evolution Canyon, a Natural Evolutionary Laboratory

The "Evolution Canyon" (EC) system in Israel turned out to be a suitable sampling place for such microevolutionary studies. Eviatar Nevo, the former director of the Institute of Evolution in Haifa, Israel, first realized the beauty and scientific potential of this natural evolutionary laboratory. EC I at Nahal Oren in the Carmel Mountains, close to Haifa, is an east-west–orientated canyon. The south-facing slope (SFS) is constantly irradiated by sun, which makes it hotter and drier, savannah-like (i.e., "Africa" like), whereas the shady "European" north-facing slope (NFS) is a mesic, lush forest (just enter "Evolution Canyon" into Google Images). The scientific attractiveness of EC is obvious. The slopes are separated by just 50 to 400 meters. Thus, geographic separation cannot account for any slope-specific intraspecies differences. The canyon system is approximately 3 million to 5 million years old and is a tectonically up-lifting canyon, until now rather undisturbed by humans. Thus, there has been sufficient time to potentially develop slope-specific intraspecies divergence. Despite having the same macroclimate (overall seasonal temperature, rainfall, etc.), the SFS and NFS have substantially different microclimatic temperature and drought stress, which allows us to identify natural selective abiotic pressures and to study their effect in nature. For more than 20 years, adaptation and speciation of macroorganisms have been explored in EC, with *Drosophila* flies and wild barley being two of the most prominent model organisms. The cyanobacterium *Nostoc linckia* and the fungi *Sordaria fimicola, Penicillium lanosum,* and *Aspergillus niger* are other model organisms.

Bacillus simplex from EC

In 2003, during my postdoc studies, I isolated a population (~1,000 strains) of *Bacillus simplex* bacteria from soil of EC I and II (located in the Upper Galilee Mountains, separated by 40 km from EC I; Fig. 1).

The first big question was, how does the diversity of the strains distribute across the two canyons (separated by 40 km), with each of the canyons having two microclimatically different slopes? Would the EC I strains be substantially different from EC II strains? This would argue for a strong influence of the geographic distance. Or would strains from different slope types be rather different, which would suggest the selective effect of different (probably abiotic) natural constraints? It turned out that strains from the same slope types (from both EC I and EC II) were very similar, despite the geographic distance, whereas strains from different slope types but within the same canyon were substantially distinct, despite their geographic proximity. This suggested that the distance of 40 km has had a rather low, if any, impact on the genetic differentiation but that the different ecologies across both slope types probably are driving the diversification.

So, we have now a first idea of how the genetic diversity of members of the same species is distributed across different canyons and slopes. But what about the genetic substructure itself? It was interesting to see that the diversity does not split into a single SFS group versus a single NFS group. Luckily, from a researcher's

Figure 1 The "Evolution Canyon" system. Left: geographical location of "Evolution Canyon" I and II in Northern Israel. Right: "Evolution Canyon" I, Nahal Oren, Haifa, Israel. Sources: http://en.wikipedia.org/wiki/File:Israel_districts.png, Google Maps; Photo right: courtesy Michael Margulis, Institute of Evolution, Haifa, Israel. doi:10.1128/9781555818470.ch31f1

perspective, the population substructure was more complex. The further analysis, now on a reduced set of ~130 strains, revealed two major phylogenetic genomic lineages, genomic lineage 1 (GL1) and GL2. As viewed simply by eye, both GL1 and GL2 contained at least one phylogenetic clade from the SFS and also from the NFS. However, to identify the biologically really relevant nodes in a phylogenetic tree at which clades split from each other is not trivial and is often subjective, as there is a multiple hierarchy of clades nested within clades. Thus, we sought to find a more objective confirmation of our subjective visual judgment. Using Ecotype Simulation, a novel algorithm developed by Frederick M. Cohan (Wesleyan University, Middletown, CT) that models the sequence diversity within a bacterial clade, we then identified phylogenetic groups (putative ecotypes [PEs]) that were objectively predicted to be ecologically and evolutionarily distinct.

But, are these PEs indeed ecologically distinct, as one would imagine from their preference for either slope type? To remind you, the strains are taxonomically very closely related. All of them are identical in their 16S ribosomal DNA sequence, which is often taken as an indication of membership in the same species. Moreover, since GL1 and GL2 each contain at least one such PE per slope type, which opens the opportunity to test microevolutionary hypothesis in parallel, would the "African" ecotypes in GL1 and GL2 express similar (adaptive) traits, despite being phylogenetically significantly diverged? Similarly, would the "European" ecotypes in both GL1 and GL2 also show similar traits?

Indeed, at the upper temperature limit of growth, the "African" strains generally grow faster than the "European" strains. Apparently, "African" strains are metabolically still active at the very high temperatures at which "European" strains already start to weaken. This was a first indication of the high-temperature adaptation of "African" strains. Further support came from the fatty acid (FA) analysis of the cell membranes. Compared to proteins and genomes, the cell membrane is hardly in the focus of mainstream research. Yet, a functioning cell membrane is crucial for microbial life. Just imagine a bottle of your favorite wine. I bet that most of us are interested only in the wine, not in the bottle itself. But once the poor disregarded bottle gets damaged, all the precious wine sadly spills out. No intact bottle, no wine; no functioning cell membrane, no living microbes. The cell membrane reacts very flexibly to changing environments, for example, by changing its FA composition. Under cold conditions, to ensure the fluidity of the membrane, bacilli incorporate predominantly anteiso-methyl-branched FAs. In contrast, iso-methyl-branched FAs are favored at hot temperatures. Interestingly, the "African" strains generally produce more iso-branched FAs than the "European" strains, irrespective of which lab temperature is used to grow the strains (different lab temperatures only test for phenotypic plasticity, i.e., the ability of the same genotype to express different phenotypes under different environmental conditions). This strongly suggests a genetically fixed difference for the preference of the branching type of incorporated temperature-relevant FAs between "African" and "European" strains, most probably as a result of a long-term exposure to different environmental temperatures.

However, not all traits follow the SFS/NFS dichotomy. Although the soil on both slopes is classified as "terra rossa," small but significant differences exist, for example, in the organic carbon content. This prompted us to test for quantitative differences in the utilization of a variety of different carbon sources. Interestingly, the observed differences were not associated with the different habitats but, rather, with the phylogenetic split into GL1 and GL2.

Future Insights

Many questions remain. For example, do "African" and "European" strains differ in their drought resistance, since the SFS soil is significantly drier? Do the spores of "African" and "European" strains differ in their heat resistance? Do the strains contain plasmids, and if so, does the abundance or diversity of plasmids correlate with either the phylogenetic structure or ecological origin of the host strains? The potential research topics appear to be endless. Currently, we are addressing some of these issues, and we expect that comparative genome sequence comparisons will soon aid us in understanding how these evolutionary changes are implemented on the molecular level of genes and their expression.

Despite the still poor amount of information, we do already understand that the observed evolutionary lineages (ecotypes) represent speciation events. These ecotypes are genetically different, reside preferentially in different habitats, and show physiological traits characteristic for the habitats in which they reside. In fact, from a biological and evolutionary perspective, they probably already represent different species, though this proposal is not supported by the current pragmatic and efficient but rather artificial taxonomic practice of microbial species delineation.

It is perhaps too naïve to expect that the observed traits indeed evolved precisely in those sites from where the *B. simplex* bacteria have been isolated, taking into account that there are probably thousands of such east-west-directed canyons on the globe and that sporulating bacteria like bacilli easily migrate with the wind from continent to continent. But most probably, such sharp microclimatic contrasts in immediate proximity reinforce within-species evolutionary splits, and therefore EC represents a beautiful site to study microevolution in natural habitats. Moreover, it illustrates the beauty of studying evolution at all!

Microbes and Evolution: The World That Darwin Never Saw
Edited by R. Kolter and S. Maloy
©2012 ASM Press, Washington, DC
doi:10.1128/9781555818470.ch32

32

On the Origin of Bacterial Pathogenic Species by Means of Natural Selection
A Tale of Coevolution

Philippe J. Sansonetti

"Nothing makes sense in biology if not in the light of evolution." One could paraphrase Theodosius Dobzansky by saying, "Nothing makes sense in the biology of bacterial pathogens if not in the light of coevolution." It is indeed most likely that this basic principle has forged pathogens and our immune system as we know them today. I recently reread Charles Darwin's *On the Origin of Species*. I realized that what had been at one time a somewhat painful student's exercise had become an intellectual delight thanks to the reflection and perspective one acquires in the course of a scientific career. Had he written his masterpiece with the knowledge of Louis Pasteur and Robert Koch's contributions, he might have selected pathogenic bacteria and their infected hosts as a model of "reciprocal natural selection." The often-forgotten second half of his book's title, *...or, the Preservation of Favored Races by the Struggle for Life,* would beautifully apply to the microbial world, showing how his basic concept transcended the models he had studied. It would

Philippe Sansonetti is Professeur at Institut Pasteur in Paris and holds the Chair of "Microbiologie et Maladies Infectieuses" at the Collège de France in Paris. He has been tracking *Shigella* throughout the gut for the past thirty years and has not yet seen the end of the tunnel.

have been a difficult task, however, because no solid phenotypic criteria for bacteria were available at the time (equivalent to the length of finch beaks). Microbes started to be identified and associated with certain diseases, but it then took several decades before scientists started to glean answers about the identity of pathogenic microorganisms in comparison to their nonpathogenic counterparts, the nature of virulence traits, and the host response—quickly termed immunity—as pioneered by Élie Metchnikoff and Paul Ehrlich at the turn of the 20th century. Diphtheria toxin (DT) was probably the first virulence factor to be identified and was subsequently purified by Emil von Behring, Émile Roux, and Alexandre Yersin. I am still amazed by the modernity of their contribution: von Behring demonstrated that sera against DT protected against the disease by neutralizing the toxic effect, avoiding the need for anti-whole bacterial cell immunity—a giant conceptual leap demonstrating that pathogenic traits could be disconnected from the microbe. Charles Nicolle got close to realizing that bacterial pathogens may have evolved from their nonpathogenic counterparts by acquiring virulence factors when he stated, "There are good reasons to believe that virulence is linked to a material support. Don't we see it sometimes undergo quick variations to which one could legitimately give the name of mutation, and these quick variations to translate, in addition to adapting to a new host, in acquiring new pathogenic properties in the animal species it ordinarily infects?" Close, but not quite there; the concept that pathogens and the immune system were the result of a long coevolutionary process that crosses species borders, even beyond the animal kingdom, had yet to come.

Genetics—more precisely, molecular genetics—was key to this understanding. The saga started with a bacterial pathogen. Frederick Griffith, while demonstrating capsule-associated reacquisition of virulence via transformation, opened the way to the discovery by Oswald Avery, Colin MacLeod, and Maclyn McCarty that DNA contained the genetic information encoding the capsule, also demonstrating that, overall, heredity was contained in DNA. Another interesting discovery from that time period was that *Corynebacterium diphtheriae* came in two "flavors": toxigenic and nontoxigenic strains. Moreover, a bacteriophage, β, was associated with expression of DT. The phage could be acquired and lost, one more step in understanding the evolution of pathogens and a second piece of evidence that virulence was acquired not so much via simple mutations but, rather, via quantum leaps involving horizontal transfer of large pieces of DNA.

This concept was soon confirmed by observing that bacteria quickly evolved in response to the expanding use of antibiotics and that a large part of this emerging resistance was acquired not so much through point mutations but through transfer of plasmids encoding enzymes that specifically hydrolyzed or modified antibiotics. *Infectious Multiple Drug Resistance,* published by Stanley Falkow in 1975, was a book that soon became my bedside reading while I was a young resident in infectious diseases and led me to Institut Pasteur to learn bacterial genetics. I realized that microbial species were not genetically frozen entities. I also measured how, under strong and recent selective pressure, such as that imposed by antibiotics, the microbial world could quickly adapt. Still, in spite of apparent acceleration, the canonic rules of evolution established a century earlier by Charles Darwin continued to be scrupulously respected thanks to the existence of pools of bacterial genes in the environment, and in our microbiota, that can be acquired via mobile genetic elements, including bacteriophages, plasmids, and transposons. These genes can be transferred into pathogenic species, sometimes establishing themselves permanently in the genome as larges pieces of DNA encoding integrated functions like adherence or invasion. This would be soon known as "pathogenicity islands" (PAI), a pioneering concept coined by Jörg Hacker in the early 1990s that is true both for gram-negative and gram-positive pathogens.

My master's degree was on the modeling of parameters of antibiotic resistance plasmid transfer in the gut. I was fascinated, but antibiotic resistance was, finally, not my cup of tea. As a clinician with a strong taste for physiology, I was fascinated by the mechanisms by which pathogens could subvert physiological mechanisms, particularly how they could cross host barriers and resist the basic mechanisms of immunity. Two diseases with which I gained experience in medical wards fascinated me by the wealth and diversity of their clinical symptoms: typhoid fever and leprosy.

Recent genomic data on the causative microorganisms have illuminated some key aspects of their pathogenesis, particularly the correlation between genome reduction and obligate intracellular parasitism in *Mycobacterium leprae*. However, one of my key questions at the time was the exquisite human specificity of these two pathogens, in spite of the promiscuous character of the *Salmonella* and *Mycobacterium* genera regarding their breadth of animal hosts.

As addressed in the last chapters of Falkow's book and soon extended, in the case of the gut flora, it had become clear that a single species like *Escherichia coli* encompassed multiple "pathovars," such as enterotoxigenic, enteropathogenic, and enteroinvasive *E. coli* (EIEC) and others, including *Shigella*. Looking back, this concept of divergent evolution generating different pathovars out of the same species caught my attention. I selected *Shigella* because I wanted to study a pathogen that caused straight invasion of an epithelium and inflammation—inflammation has always fascinated me as a medical doctor and has served as a thread in my scientific career. Thus, *Shigella* causing dysentery seemed like a smart choice, and it was so in retrospect, because it was amenable to reverse genetics, provided insight into how bacteria invade host cells, and had an extraordinary capacity to cause and regulate intestinal inflammation. The exquisite species specificity to humans and higher primates, on the other hand, as fascinating as it was, quickly turned into a nightmare caused by the lack of a fully relevant animal model.

I started to play with *Shigella sonnei*, which irreversibly loses its pathogenicity along with the expression of its somatic antigen at high frequency. It turned out to host a virulence plasmid that was lost at the frequency at which virulence was itself lost, a sort of retrograde evolution that struck me: how can *S. sonnei* survive while losing virulence at such a high frequency? Good question, but we still don't know the answer. After joining Sam Formal's group at the Walter Reed Army Institute of Research, I could confirm that a large virulence plasmid was central to the pathogenesis of all *Shigella* groups, and in a daring endeavor, we could reconstruct a fully virulent *Shigella* from *E. coli* strain K-12, by stepwise introduction of the virulence plasmid by conjugative transfer and of specified chromosomal fragments by interrupted conjugative transfer from Hfr *Shigella* strains. Not everything made sense at the time, but the in vivo model in which the successive transformants were tested was robust enough to convince us that we had probably recapitulated in acceleration, yet not necessarily in the right order, the steps of the evolutionary process that had led a gut commensal *E. coli* organism to become a *Shigella* organism. The two aspects of this genomic evolution encompassed gene addition, of course, such as the virulence plasmid, but also, unexpectedly, gene reduction. An example of the latter was the *kcp* (keratoconjunctivitis production) locus, which was a key step on the way to *Shigella* pathogenesis. Initially considered a locus that had to be added, it turned out to be a genomic "hole"—a sequence from which a lysogenic bacteriophage carrying the *ompT* gene, encoding a surface protease, had been eliminated. It would later appear that OmpT could eliminate the entire pool of plasmid-encoded, surface-expressed IcsA proteins required to nucleate cytoplasmic actin and cause cell-to-cell spread of intracellular *Shigella*. Moreover, *ompT* was functionally replaced by a virulence plasmid gene, *sopA*/*icsP*, encoding an alternative protease that preserved enough of the surface-exposed IcsA at a bacterial pole to cause polar assembly of actin in the host cell cytoplasm, thereby allowing and optimizing the motility process. This divergent evolutionary process combining subtraction and addition is a phenomenal example of the wealth of subtleties experienced by a microorganism to optimize its fitness with its host. This concept of gene deletions as part of the process of virulence optimization was nicely framed by Tony Maurelli as "black holes," or "antivirulence genes." It took, however, knowledge of the genome of isolates representative of the four *Shigella* subgroups to better appreciate the elements of plasticity that led to its current state. Regarding gene addition, several PAIs, particularly the large (40-kb) PAI present in the virulence plasmid encoding the type III secretion system of *Shigella* (TTSS) and its cognate secreted effectors, clearly differentiate *Shigella* from *E. coli* K-12. There are, however, other chromosomal additions, corresponding to PAIs of various sizes and bacteriophages. One such addition seems to possibly represent a straight mode of adaptation to acquisition of the TTSS: bacteriophage-encoded glycosylation of the somatic antigens that defines the serotypes of *Shigella flexneri* along with acetylation events and allows maintenance of similar amounts of protective somatic antigen (the shield) while imposing a helicoidal structure to the chains that reduces their length, thereby facilitating accession of TTSS tips (the sword) to epithelial cells.

Evolution of serotypes may therefore reflect a series of stochastic events (bacteriophage acquisition) that have led to serotype diversification under the double constraint of optimizing virulence while maintaining resistance to innate immune defense mechanisms, regardless of the pressure of the adaptive immune response. Regarding gene destruction, *Shigella* is characterized by a large number of pseudogenes, in comparison to *E. coli* K-12, caused either by frameshift point mutations, deletions, or integration of insertion sequences which are present in very large numbers in the *Shigella* genome, compared to *E. coli* K-12. Some of these pseudogenes qualify for the definition of "black holes," or antivirulence genes. This is the case with *cadA*, encoding lysine decarboxylase activity, which is consistently absent in *Shigella*. When production of lysine decarboxylase is restored in *Shigella*, and thus its capacity to produce cadaverine, the transformant shows a significant decrease in its invasive capacities. Moreover, further analysis indicated that when different *Shigella* and EIEC strains were compared, different types of deletions were accordingly observed. A similar observation was made with the master operon *flhDC*, which controls expression of flagella in *Shigella* and EIEC strains that are essentially nonmotile. It is likely that both cadaverine and flagellar expression are detrimental to pathogenesis; therefore, any genetic alteration inactivating these phenotypes has been selected and maintained in the course of the evolutionary reduction of the *Shigella* genome.

From these studies and many others in the field of gram-negative pathogens, it soon became obvious that the general rule of pathogen evolution is the stochastic horizontal acquisition of a genome sequence that is decisive enough, due to the new properties it encodes, to propel the microorganism into a new niche to which it is not adapted. Under this struggle-for-life condition, only those microorganisms that can adapt survive. This was predicted by Darwin: "It is not the strongest of the species that survives…nor the most intelligent that survives. It is the one that is the most adaptable to change." So are forged pathovars, through further events of gene addition or reduction, topped by a hierarchy of regulations themselves networked with a complex array of two-component sensors.

In the prokaryote world, pathogenesis often stems from addition of genes, looking like a true arms race with the host, or sometimes from massive reduction that follows the increased adaptation to a niche imposed by previous gene acquisition. The latter is particularly seen for microorganisms that have become obligate intracellular parasites (e.g., *Chlamydia*). Whether gene loss has preceded and driven evolution to this absolute state of dependence, or whether genes have been lost following adaptation, as matter of "stealth" (i.e., "black holes"), or are due to a lack of selective pressure in the new niche that provides to most metabolic needs, remains an enigma. Pushed to an extreme, those microbes may eventually become true endosymbionts, like *Buchnera,* which has undergone massive genome reduction (down to 0.64 Mb). Almost all of its 590 genes have close homologs in enteric bacteria, indicating, along with phylogenetic data, that it derived from a much larger genome resembling modern enteric bacteria such as *E. coli*.

In the case of *Shigella,* if it is clear that the decisive event was the acquisition of the virulence plasmid by a commensal strain of *E. coli,* what is not clear is whether the event occurred several times in different *E. coli* isolates belonging to different phylogenic groups, or if the plasmid transfer event occurred once and, from there, all groups and serotypes evolved from the same ancestor. More recently, it has become evident that some TTSS-secreted effector proteins (i.e., Osps and IpaHs) were enzymes that attacked "head on" the signaling cascades of the innate defense of host cells, such as the NF-κB and mitogen-activated protein kinase pathway. Elaborate enzymatic activities such as deubiquitinases and phospho-threonine lyases, an original family of E3 ligases, have been identified in both animal and plant pathogens. These sequences have in general poor, if any, homology with sequences of the genes encoding equivalent enzymatic activities in eukaryotes. Functional convergence has probably been the force behind the appearance of these enzymes. However, it remains a mystery in which original species (eubacteria, archaea, or primitive eukarya like protozoans) the core of genes that led to the diversification of the current pool of TTSS effectors originally appeared. I like to think that this core was assembled in deep-seawater ancestor microorganisms fighting planktonic protozoans, before the primitive bacterial world became terrestrial. After all, as predators, what is the difference between an amoeba and a macrophage?

We may find some *déjà vu* in this evolving world of pathogenic prokaryotes which have often found solutions at the cost of minimal adaptation of this core gene pool, seen as an endlessly adaptable toolbox. It is likely that the species that "forged" this core gene pool have disappeared. Still, how did bacteria acquire ubiquitine ligases and deubiquitinases, since they themselves do not produce or use ubiquitine? Similarly, the making of a TTSS out of a flagellum, or vice versa, is another example showing that under massive selective pressure, recycling the existing components is more efficient than creating novel solutions—which is a blow to creationists, and not necessarily a compliment to nature's inventive capacity. This was beautifully phrased by François Jacob: "It is natural selection that gives direction to changes, orients chance, and slowly, progressively produces more complex structures, new organs, and new species. Novelties come from previously unseen association of old material. To create is to recombine."

Let's leave the last word to Charles Darwin: "Nature is prolific in diversity but niggard in innovation."

Microbes and Evolution: The World That Darwin Never Saw
Edited by R. Kolter and S. Maloy

doi:10.1128/9781555818470.ch33

33

The Evolution of Diversity and the Emergence of Rules Governing Phenotypic Evolution

Paul B. Rainey

Charles Darwin was fascinated with life's diversity. His fascination began when he was a young child, but it was fueled by serendipitous associations during his university education with leading intellects that opened his mind to the wonders of the natural world. In December 1831, a young Darwin joined the second survey expedition of HMS *Beagle* under the command of Robert Fitz Roy as a naturalist and gentleman companion to the captain. During the course of the 5-year voyage, Darwin came face-to-face with life's bewildering diversity—in a way that few had previously experienced. Unlike contemporary natural historians who saw evidence of divine design in nature and explained adaptations as God acting through the laws of nature, Darwin saw reason to question: he was puzzled by the geographical distribution of species and the fossils

Paul Rainey completed his Ph.D. at the University of Canterbury in New Zealand and did postdoctoral research at the University of Cambridge and then at the University of Oxford. He joined the faculty at Oxford in 1994. Since 2007 he has been at the New Zealand Institute for Advanced Study in Auckland. He is also a Member of the Max Planck Society and holds an adjunct position as Honorary Director at the Max Planck Institute for Evolutionary Biology in Plön, Germany. He spends as much time as possible at his hut in Cecelia Sudden Bay on remote Great Barrier Island, where he cannot be reached.

he collected. His puzzlement led him to consider the possibility that species were not fixed immutable entities placed on Earth by a divine being but might in fact alter—one species into another. Ultimately, such ideas led Darwin to propose a mechanism to explain the diversity of life, namely, evolution by natural selection.

Of the many places Darwin visited on the voyage of the *Beagle,* none was more influential than the Galapagos Islands. In his account of the natural history and geology of the voyage, Darwin reports being struck by the degree of endemism of the flora and fauna evident even at the level of different islands of the archipelago. At the same time, he notes a strong relationship to the South American flora and fauna and hints at the possibility that a few stray colonists from the mainland may have been ancestral forms of those in the Galapagos. Darwin then proceeds to describe differences among species collected in the islands, focusing particularly on "a most singular group of finches, related to each other in the structure of their beaks, short tails, form of body, and plumage" (Darwin, 1845, p. 379). Darwin is especially curious about the gradation of beak sizes and their corresponding uses and famously notes (p. 380), "Seeing this gradation and diversity of structure in one small, intimately related group of birds, one might really fancy that from an original paucity of birds in this archipelago, one species had been taken and modified for different ends."

In all but name, Darwin described an evolutionary process involving the splitting of a single lineage into a number of niche-adapted types, a process that since the early 1950s has become known as adaptive radiation. Adaptive radiations are in fact commonplace; indeed, the vast majority of life's diversity appears to have arisen through successive adaptive radiations, the most famous perhaps being the radiation that took place approximately 530 million years ago during the Cambrian period (the Cambrian radiation) and from which arose the major groups of complex animals.

While evidence of adaptive radiation abounds and is supported by a solid theoretical framework, much insight into the moment-by-moment workings of adaptive radiation has come from laboratory studies with microbial populations. My own interest in the emergence of diversity began with the common plant- and soil-colonizing bacterium *Pseudomonas fluorescens* and its propagation in unshaken test tubes (microcosms). Daily sampling of evolving populations shows the rapid emergence of diversity, as evident by a range of morphologically distinct colony variants on agar plates (Fig. 1).

Striking is the fact that different morphotypes, when placed back into pristine microcosms, show evidence of niche specificity. In essence, the diversification of *Pseudomonas* populations in static broth microcosms is no different from the radiation of finches on the Galapagos Islands as envisaged by Darwin. But unlike the finch radiation, the *Pseudomonas* radiation can be run, and rerun, from identical starting positions as often as the experimenter wishes—the experimenter being able to watch the process of adaptive radiation in real time. In addition, the serendipitous correspondence between colony morphology and niche preference means that the process of adaptive radiation can be studied by simply scoring the change in frequency of colony morphologies through time.

The model *Pseudomonas* radiation has been used to test a number of ideas relating to the origin and maintenance of life's diversity, but in the context of adaptive radiations, two predictions from ecological and evolutionary theory hold center stage, namely, that diversification depends upon ecological opportunity (vacant niche space) and that competition for limiting resources fuels the process of diversification.

The first prediction was tested by allowing replicate populations to evolve in either shaken or unshaken microcosms: microcosms incubated without shaking are replete with ecological opportunity, whereas microcosms incubated with shaking are depleted in ecological opportunity. Consistent with theory, diversity emerged in the static environment but not in the shaken environment. Further experimental work revealed evidence of trade-offs among the niche specialist genotypes, which left little doubt that competition was the engine driving the *Pseudomonas* radiation.

Since its initial use to explore the validity of theory surrounding adaptive radiation, the experimental *Pseudomonas* radiation has been used extensively to obtain insight into a wide range of

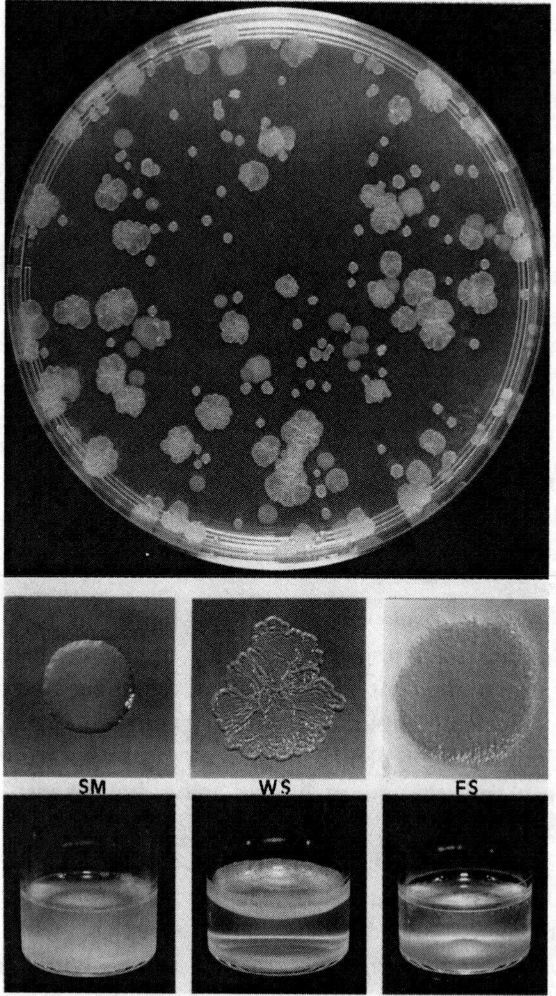

Figure 1 Adaptive radiation in a test tube. Populations were founded by a single ancestral "smooth" (SM) genotype of *P. fluorescens* and propagated in 6 ml of King's medium B. Microcosms were incubated without shaking to produce a spatially structured environment replete with ecological opportunity. After 7 days, populations show substantial phenotypic diversity, evident upon plating of samples onto agar plates (top). The bewildering array of morphotypes can be assigned to a number of different classes: SM, wrinkly spreader (WS), and fuzzy spreader (FS) (middle). The derived genotypes show striking evidence of niche specialization (bottom). From Rainey and Travisano (1998). doi:10.1128/9781555818470.ch33f1

evolutionary processes. While the questions tackled have been many and various, my own interest has been influenced by a desire to understand the moment-by-moment workings of evolution. While Darwin provided a sound theoretical framework for the study of evolution—including a mechanism to explain the fit between organisms and environment—he had little idea of the mechanistic basis. Necessary to bring the study of evolution into line with advances in genetics and molecular biology is knowledge of the mutational origins of new phenotypes, and, critically, understanding of the connection between genotype and phenotype in sufficient detail so as to explain precisely why natural selection favors one type over another.

As mechanistic insight into the *Pseudomonas* radiation has been gained, my interest in the repeatability of evolution has been stimulated. There has been a long-standing interest in this question: in his book on the Cambrian radiation, Stephen J. Gould suggested the Gedankenexperiment of replaying life's tape to test the repeatability of evolution. While experiments on the grand scale that Gould envisaged (replaying the Cambrian explosion!) are not possible, experiments with microbes have allowed the roles of chance, adaptation, and history in evolution to be appraised. Perhaps surprisingly, these experiments have revealed that when selection is strong, evolution can be remarkably repeatable. Indeed, in our own experiments, the adaptive radiation that takes place in static broth microcosms is highly repeatable. The most common type, the wrinkly spreader—of which there are numerous subtly different forms—arises reliably after 2 or 3 days from the ancestral genotype and follows highly similar dynamics in independent microcosms (Rainey and Travisano, 1998).

Most would attribute the repeated emergence of the wrinkly spreader type to strong selection, large population sizes, and, no doubt, genetic uniformity of the starting material. Indeed, experiments in which population sizes are reduced (and thus, the likelihood that a given population will harbor a cell with a mutation causing the wrinkly spreader form is reduced) show that the reproducibility apparent when population sizes are on the order of 1 billion cells per milliliter is absent when population sizes are reduced by 2 orders of magnitude. But is this the end of the story, or could it be that evolution follows particular paths? Are there properties of genetic architecture—the way organisms are wired, the particular functional and regulatory connections that link genotype to phenotype—that bias the outcome of evolution? Some think so; in fact, scientists interested in evolution and development have long argued that "developmental constraints" are likely to influence the outcome of evolution (Maynard Smith et al., 1985).

The recent discovery of repeated molecular evolution—not just in laboratory populations of microbes but also in natural populations of insects and plants—has brought renewed interest to this possibility, but mechanistic insight of the hard-proof variety has been lacking. Nonetheless, as the tools of molecular biology become ever more powerful, it is just possible that the relative contributions of genetic architecture and natural selection to evolutionary change can be disentangled. Indeed, our recent work on the genetics of wrinkly spreader diversity provides evidence that genetic architecture does indeed contribute to the repeatability of evolution by biasing the molecular variation presented to selection.

Painstaking genetic and molecular analyses performed over many years have shown that single (simple) mutations are alone sufficient to generate a diverse collection of wrinkly spreader forms, but interestingly, these mutations are clustered in just three loci: *wsp*, *aws*, and *mws*. While the three loci are distinct, each contains a gene encoding a diguanylate cyclase (DGC). In each instance, the wrinkly spreader-causing mutation destroys the function of the regulator that suppresses activation of the DGC. This results in constitutive activation of the DGC and overproduction of the secondary signaling molecule cyclic di-GMP. In turn, overproduction of cyclic di-GMP causes constitutive activation of enzymes that produce adhesive cellulose-based polymers, the primary structural determinant of the wrinkly spreader form (Fig. 2).

Curiously, the genome of *P. fluorescens* contains 39 DGCs—any one of which in principle could, if overactivated, generate the wrinkly spreader form. Why, then, are all naturally occurring wrinkly spreaders the result of mutations in one of just three (*wsp*, *aws*, and *mws*) pathways? Is this because the additional 36 DGCs

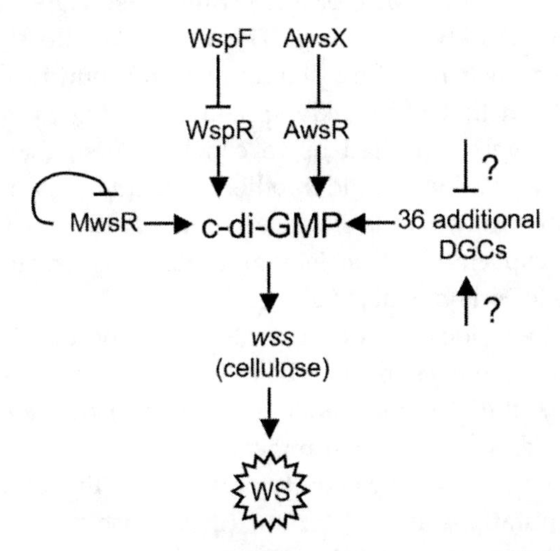

Figure 2 Network diagram of DGC-encoding pathways underpinning evolution of the wrinkly spreader (WS) phenotype and their regulation. Overproduction of cyclic di-GMP (c-di-GMP) results in overproduction of cellulose and other adhesive factors that determine the WS phenotype. The ancestral SBW25 genome contains 39 putative DGCs, each in principle capable of synthesizing the production of c-di-GMP, and yet WS genotypes arise most commonly as a consequence of mutations in just three DGC-containing pathways: Wsp, Aws, and Mws. In each instance the causal mutations are most commonly in the negative regulatory component: *wspF*, *awsX*, and the phosphodiesterase domain of *mwsR*. From McDonald et al., 2009. doi:10.1128/9781555818470.ch33f2

cannot be activated by single mutations, or is there some bias inherent in the *wsp*, *aws*, and *mws* pathways that ensures that natural selection is far more likely to see mutations in these pathways relative to the remaining pathways?

To address this question, a derivative of the ancestral genotype from which the *wsp*, *aws*, and *mws* loci were eliminated was genetically engineered. This genotype was then allowed to evolve in a static broth microcosm, and derived populations were examined for the presence of wrinkly spreader colonies. To our surprise, such colonies were found, but instead of appearing at day 2, they took another 3 days to arise. The increased time to detection of these "slow wrinkly spreaders" could reflect one of two factors: a severely restricted spectrum of mutational possibilities to generate the wrinkly spreader form, or a low fitness associated with the so-called slow wrinkly spreaders. The fitness of the slow wrinkly spreader type was therefore determined and found to be indistinguishable from that of wrinkly spreaders arising by mutations in *wsp*, *aws*, or *mws*. This led us to conclude first, that mutational pathways to wrinkly spreader other than *wsp*, *aws*, and *mws* do exist (but are rarely trodden) and, second, that these pathways have a reduced capacity to translate mutation into wrinkly spreader variation (McDonald et al., 2009).

Why, then, does evolution rarely ever proceed via a non-*wsp*, non-*aws*, or non-*mws* route? Unifying the *wsp*, *aws*, and *mws* pathways is the presence of a DGC, but also negative regulation—the presence of a component of the pathway that, when functional, suppresses the activity of the DGC (Fig. 2). Recognizing that most mutations are deleterious (they cause a loss of function), any DGC-containing module subject to negative regulation will translate DNA sequence variation (mutation) into wrinkly spreader variation at a high rate, because any mutation that abolishes function of the negative regulator will result in activation of the DGC. Contrast this with a DGC-containing pathway in which regulation is via a positive activator: relatively few mutations will generate the activate state of the regulator. On a per-locus basis, then, a DGC module involving negative regulation will have a greater propensity to produce wrinkly spreader variation relative to a pathway involving positive regulation. It follows, then, that the spectrum of mutations generating phenotypic variation presented to natural selection will be a biased subset of all possible variation.

Darwin's insight into evolution and his formulation of a mechanism (natural selection) to account for life's diversity relied upon a combination of observation, theory, and experiment, not to mention a critical and inquiring mind. While evidence in support of Darwin's theory mounts daily, there is much yet to be unraveled by way of mechanistic detail. Not only does such detail provide substantive insight into the process of evolution, but also it fuels the discovery of new problems and challenges that spur the study of evolution further onward. Among the most exciting of contemporary issues is genetic architecture—the genotype-to-phenotype map—and its influence on the evolution of populations. With the emergence of improved understanding of this map—its definition, its formulation, and the constraints it imposes—rules by which the outcome of phenotypic evolution might be predicted may just arise. From a personal perspective, one of the most fascinating questions concerns the modular nature of regulatory systems, which would appear to impart fuel to evolution (some would say evolvability). Could it be that such regulatory arrangements evolved because they fuel evolutionary change—such as is necessary for a lineage to diversify—or are they simply a consequence of the way organisms need to be wired in order to function as individual entities? Without a doubt, studies of evolutionary process using microbes will provide insight.

FURTHER READING

Darwin C. 1845. *Journal of Researches into the Natural History and Geology of the Countries Visited During the Voyage of H.M.S.* Beagle *Round the World*, 2nd ed. John Murray, London, United Kingdom.

Gould SJ. 1989. *Wonderful Life: The Burgess Shale and the Nature of History.* Penguin, London, United Kingdom.

Maynard Smith J, Burian R, Kauffman S, Alberch P, Campbell J, Goodwin B, Lande R, Raup D, Wolpert L. 1985. Developmental constraints and evolution. *Q Rev Biol* **60:**265–287.

McDonald MJ, Gehrig SM, Meintjes PL, Zhang XX, Rainey PB. 2009. Adaptive divergence in experimental populations of *Pseudomonas fluorescens*. IV. Genetic constraints guide evolutionary trajectories in a parallel adaptive radiation. *Genetics* **183:**1041–1053.

Rainey PB, Travisano M. 1998. Adaptive radiation in a heterogeneous environment. *Nature* **394:**69–72.

Microbes and Evolution: The World That Darwin Never Saw
Edited by R. Kolter and S. Maloy

doi:10.1128/9781555818470.ch34

34

The Christmas Fungus on Christmas Island

Anne Pringle

The fungus *Amanita muscaria* has a mushroom that looks startlingly familiar whether it is collected from France, Russia, Alaska, California, or New Zealand; the mushroom is bright red and has white spots. This is the species most often targeted by the fairies and bunnies drawn for children's books. It is the species that Tintin discovered as his "L'Étoile Mystérieuse," and it is also the mushroom that killed the king of the elephants in the original Babar story. *A. muscaria* is a famous fungus, and we seem to find it all over the world.

Although fungi collected from different parts of the world may look the same, traveling around the world is not so easy, even for a fungus. The mushrooms collected in France and California are not the same species. They look alike, but the genetic data biologists now use to delineate patterns of ancestry among different populations identify the *A. muscaria* isolates of different continents as different species.

Despite the formal taxonomy, people around the world continue to recognize this species complex as the home of fairies and also as a traditional decoration for Christmas trees. In fact, various

Anne Pringle learned mycology at the University of Chicago, Duke University, and University of California, Berkeley. She joined the faculty at Harvard University in 2005. Although she's never traveled to Christmas Island, she'd happily go there or just about anywhere else in the universe.

writers have suggested that *A. muscaria* was the inspiration for Santa Claus. The Eurasian species is hallucinogenic. Stories of Siberian shamans using *A. muscaria* to reach the spirit world and carry back presents caused the poet Robert Graves (among others) to speculate about connections between Father Christmas and the shamanistic traditions of Siberia. After all, both St. Nicholas and the mushroom dress in red, are trimmed with white, and fly or enable flight through Northern climes. Shamans returned from the spirit world by climbing through the smoke holes of yurt ceilings. Siberia is also the home of reindeer. And although the mythology is probably wrong, you will still hear the story of Santa Claus and the mushroom retold by many people.

Christmas Island is a remote dot in the middle of the Indian Ocean, 500 km south of the Indonesian capital of Jakarta. A territory of Australia today, it was uninhabited until the late 19th century. Covered in rainforests, over half of the island is currently protected as a national park. From my library in Cambridge, Massachusetts, I cannot access information about the fungi of Christmas Island, although searches on the web give a lot of pages for a set of stamps illustrated with mushrooms. Does the Christmas fungus grow on Christmas Island? I can't tell you. And of course that's a common state of affairs in fungal or microbial biogeography; often we don't know species' ranges.

Answering a question about a species in a specific habitat depends on understanding whether or not it has arrived at the habitat and whether the habitat provides all that is necessary for the species to grow. Dispersal is a key control on the biogeography of organisms. The assumption that fungal spores are passively dispersed by air and water is at odds with obvious biomechanical adaptations to reach or create wind. These include the catapult of ballistospores and the explosive launch of ascopores. Fungi do, in fact, control many aspects of their dispersal. Even so, there are very few data on how far a fungal spore will normally travel; it's not clear if *A. muscaria* would be able to reach Christmas Island from the places where we know it does grow. As to whether Christmas Island provides a good habitat, the fungus grows as a symbiont of plants, and if compatible hosts are around, probably it would grow. But the last and most poignant answer to questions about the current biogeographies of fungi involves humans and our ability to move species across the planet.

In nature, fungal dispersal may normally involve active dispersal, as well as wind and water, but now humans also carry species among continents. Humans move species by accident and also to support agriculture, forestry, and horticulture. Because many trees cannot grow unless they are planted with symbiotic fungi, species like *A. muscaria* are used by foresters to inoculate seedlings planted for timber. The fungi function as extensions of a root system and facilitate access to scarce soil resources. For this reason, tins of soil with fungal spores were shipped from Europe and across southern Africa. Members of the *A. muscaria* species complex were introduced to Africa and also Australia, New Zealand, Hawaii, and South America. In Australia and New Zealand, the species is spreading. And so if the Christmas fungus doesn't grow on Christmas Island now, it may eventually reach it, carried by humans. If there are local hosts, the species will invade.

Darwin observed that "barriers of any kind, or obstacles to free migration, are related in a close and important manner to the differences between the productions of various regions" (p. 347; Costa, 2009). By productions, Darwin meant species. There are probably at least one million species of fungi on Earth, but fewer than 100,000 of these have been isolated, identified, and given a formal name. Moreover, there are not enough data to predict how the diversity of fungi is distributed across the Earth. For example, we do not know if fungi follow Wallace's line, so that an entirely different group of species would be found on Christmas Island and Bali (to the west of Wallace's line) than on Lombok and in the rest of Australasia (to the east of Wallace's line). Meanwhile, humans are rapidly moving them among continents, circumventing barriers and obstacles to free migration. Although Darwin published his book just over 150 years ago, in that short time humans have radically reshaped the natural history of our world. The answer to a question about the biogeography of *A. muscaria* was different when Darwin was alive, as compared to now, and it will probably be very different in another 150 years.

Microbes are rarely the targets of conservation, but data for plants and animals suggest that introduced species cause harm to local biodiversity. Extinctions are associated with introduced species that spread and fill a habitat. Fungi may go unnamed and ignored, but it seems quite likely that climate change, habitat loss, and also introduced species will cause both a massive rearrangement of current fungal biodiversity and extinctions.

In the end, invasions and extinctions provide both challenges and an opportunity. One challenge is to disentangle the forces threatening species and provide tools for conservation. On the other hand, recording and interpreting patterns within the novel communities created by humans provide an opportunity to elucidate the evolution of microbial species. If the Christmas fungus arrives for a first time on Christmas Island, what will it do there?

FURTHER READING

Costa JT. 2009. *The Annotated* Origin*: A Facsimile of the First Edition of* On the Origin of Species, *by C. Darwin*. Belknap Press of Harvard University Press, Cambridge, MA.

Letcher A. 2007. *Shroom: A Cultural History of the Magic Mushroom.* HarperCollins, New York, NY.

Vellinga EC, Wolfe BE, Pringle A. 2009. Global patterns of ectomycorrhizal introductions. *New Phytol* **181:**960–973.

Microbes and Evolution: The World That Darwin Never Saw
Edited by R. Kolter and S. Maloy

doi:10.1128/9781555818470.ch35

35

A New Age of Naturalists

Rachel J. Whitaker

Charles Darwin and other naturalists of his time traversed the globe to survey the diversity of plants, animals, and insects and their distributions around the world. They combined these surveys with their contemporary understanding about the history of Earth to develop an evolutionary model for the origins of species through natural selection. In Darwin's *On the Origin of Species*, the geographical distribution of plants and animals plays a central role in his developing the theory of descent with modification. Darwin and others noticed that organisms from the "same areas of land and water" shared characteristics which he called the "deep organic bond" of inheritance. Darwin argues, based on this bond, that "varieties" of organisms must have descended from a common ancestor that dispersed and later became disconnected and diverged. He suggests a mechanism of divergence as, "dissimilarities of the inhabitants of different regions may be attributed to modification through variation and natural selection." To back up this argument, two chapters of *Origin* are devoted to developing plausible dispersal mechanisms and changes in landscape that could link to common ancestor populations that now appear to be disconnected.

Rachel Whitaker did her doctoral and postdoctoral research at the University of California, Berkeley. She is an assistant professor in the Department of Microbiology at the University of Illinois at Urbana-Champaign. She likes nothing more than outdoor adventures with her husband and two young boys.

One hundred fifty years later, microbial naturalists are uncovering a whole new world of diversity that will lead to a sea change in our understanding of evolution. Using DNA sequencing as a powerful lens through which to resolve relationships between minute cells that look the same even with the most powerful microscopes, we are discovering new microbes each year, even in the most mundane places. Already, the diversity of microbes that these tools have uncovered puts the entire menagerie of macroorganisms to shame and has changed the way we understand the evolution of life on Earth. Using DNA sequences to map the relationships between these and other organisms has resulted in a completely new tree of life on which humans, and the vast array of plants, animals, insects, and fungi that captivated Darwin, are just tiny twigs. The new tree shows that we are not at the top of the evolutionary ladder but are equal, in terms of evolutionary history, to all of the extant microbes on Earth today. Most spectacularly, reordering these relationships revealed three domains of life—*Bacteria, Archaea,* and *Eukarya*—instead of the five kingdoms most of us grew up studying.

It was the discovery of the *Archaea,* gaining broad acceptance in the early 1990s, that motivated me to become a microbiologist. I was inspired by the 1997 issue of the journal *Science* on frontiers in microbial biology, in which Norman Pace described the microbial diversity that was being uncovered by the new, high-resolution molecular lens. A companion article described the extraordinary work of Carl Woese, who had pioneered the field using molecular tools to lay out the framework for the tree of life, had identified *Archaea,* and was continuing to investigate the origins of life and the fundamentals of cellular evolution. How could I resist the chance to study a new domain of life? The challenge for me, and for the next generation of microbial naturalists as the breadth of diversity continues to be uncovered, is to understand how this diversity evolved. Are the evolutionary mechanisms of microbes similar to those of their distant macrobial cousins? Many of the answers will come, as they did for Darwin and the naturalists of his time, by looking at the natural distribution of microbial diversity and putting it in its appropriate context of space and time.

Evolutionary biologists studying macroorganisms have been studying the mechanisms through which diversity evolves—the origins of species—for hundreds of years. Conventional evolutionary theory says that new species arise when populations become disconnected so that DNA is unable to move between them. What constitutes a genetic barrier that can disconnect populations? For macroorganisms, the primary mechanism is geographic isolation, such as a mountain range or an ocean, that prevents migration and consequent mixing between populations. These are the barriers that Darwin discusses at length in *Origin*. Once populations are isolated, they are free to adapt specifically to the local environmental conditions. Natural selection can favor a long, pointed beak in one place and a blunt, rounded beak in another. These barriers therefore allow "dissimilarities" or diversity to evolve. Of course, evolution does not only occur through natural selection; accidents can cause the extinction of one population and the continuation of another. The key is that even if random events drive divergence, once isolated, populations are independent and free to follow their own evolutionary course.

Initial surveys of microbial diversity suggested that microbes did not fit this conventional model of evolution. Microbial naturalists were unable to identify mechanisms that would disconnect microbial populations from one another. Microbes seemed too small and too abundant to be restricted by typical geographic barriers. In addition, the lens of DNA sequencing of highly conserved genes uncovered seemingly identical organisms in far-flung corners of the globe, suggesting that geographic isolation did not exist for microorganisms.

Struck by the conundrum of extreme diversity with no evolutionary mechanism to promote it, John Taylor and I set out to determine whether geographic barriers could isolate populations and allow their independent evolution. To test this, we turned to the so-called "extremophiles," which prefer what to humans are extreme conditions and thus are the least likely to be able to travel the world freely. We chose to examine *Sulfolobus*, from the domain *Archaea*, which prefers volcanic hot springs where the water is close to the pH of battery acid at a scalding 80°C. We reasoned that with such a limited and discontinuous habitat, if geographic barriers existed for any microorganism, we would find them there. Based upon the increasing resolving power provided by the lens of complete genome sequences, we set out to test whether *Sulfolobus* would be able to disperse across the globe.

This is where the adventuresome part of being a microbial naturalist comes in. To collect samples from remote locations, we got to helicopter into the caldera of a volcano, travel by Soviet-era troop transport over roadless tracts of the Kamchatka Peninsula, and bushwhack into trailless areas of Yellowstone and Lassen national parks to find hot-spring basins untouched by tourists. Back in the lab, these adventures paid off when we found that each population was endemic to each geothermal region. *Sulfolobus* cannot successfully disperse across large distances. Although not visible under a microscope, the sequence differences between *Sulfolobus* isolates from distant locations are in the same range as the differences between Darwin's finches on the Galapagos Islands and the species on the mainland, which are clearly discernible by their physical features.

Biogeographic distributions of microbial species have now been described for many environments that are not considered extreme, such as soil, human intestines, and freshwater lakes. Even environmentally similar locations that are separated by small distances, in some cases, appear to host unique microbial populations. If biogeographic patterns for microbes become the rule, microbial diversity is orders of magnitude greater than our already eye-popping recent estimates. Imagine the vastness of microbial diversity in which unique types are endemic to each centimeter of soil on Earth!

Beyond uncovering diversity of literally astronomical proportions, the biogeographic distribution of microbes opens a new window on the evolutionary processes that shape each one. In line with the theory of descent with modification, each microbial population can adapt to its unique local "island" habitat, allowing microbiologists to link an organism to its environment to uncover essential drivers of natural selection for different traits. Independent populations provide replicates in nature's experiment under different conditions. The challenge is to resolve the difference in conditions on a microbial scale. In addition, physically isolated populations are much more likely than globally distributed populations to acquire differences that result from random chance events. Fragmented microbial populations are more prone to extinctions. This could change how we view microbial diversity. Instead of assuming that all variation is functionally important to a cell's survival, geographic isolation allows variation to exist for no reason other than historical accident.

Already, what has been found by examining the evolutionary process in island microbial populations has surprised us. The system of isolated volcanic "islands" hosting unique but similar species of *Sulfolobus* allowed us to trace the independent evolutionary history of three populations from a shared common ancestor. Again, using high-resolution genome sequencing, we identified the elements of genomes that had changed since populations diverged.

Our initial analyses suggest that primary differences come not from natural selection by abiotic environmental variables between locations, but from interactions between microbes in each location, specifically between *Sulfolobus* and its viruses and other microbial parasites. Since viruses are not only predators but also agents of gene transfer, moving DNA between often divergent cells, it appears that interactions between *Sulfolobus* cells and their viruses play a large part in defining the diversity and evolutionary history of *Sulfolobus*. This is something that is likely to distinguish the evolutionary process of microorganisms from that of their distant macrobial cousins.

Sulfolobus populations may be disconnected from each other, but each one is tapped into a wealth of genetic potential essentially "housed" in viral storage. The factors that define the spread of genetic material through this viral conduit are still unknown; however, investigating these dynamics uncovered even more surprises. Recently, elegant work by many others in the field identified an adaptive immune system that plays an important role in the interactions between many microorganisms and their parasites in natural environments. Studying the evolution of this newly discovered immune system in *Sulfolobus* provides an even stronger lens into the evolutionary dynamics within microbial populations. Based on the mechanisms that have been established for this adaptive immune system, it appears that this system evolves through a process that is more Lamarkian than Darwinian; i.e., adaptive changes within the genome of *Sulfolobus* occur not through random undirected mutation but in response to viral infection. The impact of these dynamics is only beginning to be understood. Watching evolutionary dynamics unfold in natural populations like those we have defined for *Sulfolobus* demonstrates the power of looking outside the lab at the evolutionary process happening all around us.

Despite the intrigue, adventure, and potential for discovery, it seems to me that microbial naturalism has not caught on as much as it should have in the 21st century. The study of microbial evolution lags behind that of macroorganisms, and it seems that there are relatively few microbiologists choosing microbial naturalism. Historically, this was because we lacked the strong lens we have now with sequencing tools that have both revealed microbial diversity and allowed us to develop new evolutionary models to understand it. Rather than exploring the distribution of diversity in the natural world, most microbiologists instead choose a very different approach to microbiology, using genetic and molecular biology in laboratory experiments to disassemble the inner workings of a few captive model microbes.

When I entered graduate school, I faced a tough decision between the two historically alienated conceptual frameworks of microbiology. I was tempted by the approaches of genetics and molecular biology, which solve complex puzzles in a way that is satisfying, elegant, and precise. I was most interested in microbial evolution, but letting nature conduct the experiments and learning to interpret the results seemed so much more complex and diffuse, often needing theoretical modeling and statistics. Ultimately, I was swayed by the compelling questions of microbial evolution. I was fascinated not just by the mechanics of the cellular machinery but also by the prospect of learning how the diversity of these machines came to be.

I still perceive the tension between these two approaches to microbiology, which has recently been eloquently described by Carl Woese and Nigel Goldenfeld. However, more and more the power of a meaningful synthesis of the two approaches is being demonstrated. After 50 years of molecular biology, we know more about biology than Darwin could have dreamed. Integrating this knowledge to understand the distribution of diversity will be the basis for microbial naturalism and will allow microbiologists to develop a deeper understanding of diversity than has ever been attained. It is clear that the new generation of microbial naturalists who set out, as Darwin did, to explore the distribution of natural diversity now wields not only the strong lens of genomics but also the power of molecular biology and genetics, which will enable them to discover the fundamental laws of evolution that apply to all of life on Earth.

I thank Carl Woese, Norman Pace, John Taylor, and Jillian Banfield for inspiration and guidance. I thank Dennis Grogan for introducing me to Sulfolobus *and his dedication to applying the power of genetics to this nonmodel organism. In addition, I thank lab members and colleagues at the University of Illinois; the National Science Foundation and NASA for funding; and Stephen Wald, Nicole Held, and Carin Vanderpool for comments on making the manuscript more accessible to a general audience.*

FURTHER READING

Horvath P, Barrangou R. 2010. CRISPR/Cas, the immune system of bacteria and archaea. *Science* **327:**167–170.
Morell V. 1997. Microbial biology: microbiology's scarred revolutionary. *Science* **276:**699–702.
Pace N. 1997. A molecular view of microbial diversity and the biosphere. *Science* **276:**730–734.
Whitaker R. 2006. Allopatric origins of microbial species. *Phil Trans R Soc B* **361:**1975–1984.
Woese CR, Goldenfeld N. 2009. How the microbial world saved evolution from the Scylla of molecular biology and the Charybdis of the modern synthesis. *Microbiol Mol Biol Rev* **73:**14–21.

Microbes and Evolution: The World That Darwin Never Saw
Edited by R. Kolter and S. Maloy

doi:10.1128/9781555818470.ch36

36

The Ship That Led to Shape

Kevin D. Young

> There is grandeur in this view of life,...[that] from so simple a beginning endless forms most beautiful and most wonderful have been, and are being, evolved.
>
> Charles Darwin, *On the Origin of Species*

Please don't tell my mother, but for years I've wanted to *be* Charles Darwin.

Not all my life, of course, and not because of his book *On the Origin of Species*, which we are here commemorating, and not because I have anything against my mother, but because of the book we now know as *The Voyage of the* Beagle. I read Darwin's extended journal well after I was embedded in my own scientific career, and what I came away with was envy. I like to think of it as a justifiable, virtuous envy: not the base desire to deny joy to someone else, but an envy that, according to Aristotle, manifests as a "pain caused by the good fortune of others." I could hardly read a page without thinking that what Darwin experienced was both magnificent and beyond reach, an unrepeatable personal and scientific adventure. It was, as David Amigoni describes the

Kevin Young received his Ph.D. from the University of Oklahoma, followed by postdoctoral stints at Texas A&M and UC Berkeley. He spent 24 winters at the University of North Dakota School of Medicine and in 2009 moved to the University of Arkansas for Medical Sciences. He tries to read widely, attempts to write poetry, and sings when he thinks no one can hear (and sometimes when they can).

narrative, "a brilliant evocation of an awe-struck encounter with the natural world in all its variety." My feelings exactly.

The Voyage of the Beagle made me long to be with Darwin, or take his place. To travel the world when much was unknown, mysterious and untouched, or very nearly so, and accessible to the sharp-eyed without encumbering equipment. To share his tortuous ascent of the mountain in Tierra del Fuego, where he felt that "Death, instead of Life, seemed the predominant spirit," but on reaching the top was struck by the "grandeur of the scene" (*Darwin: The Voyage of the* Beagle, p. 200 [Griffith, 1997]). Or to stand on that crest in southern Chile which sparked in him emotions as profound as those accompanying "watching a thunderstorm, or hearing in full orchestra a chorus of the *Messiah*" (p. 307). Or to walk "the primeval forests undefaced by the hand of man" and "feel that there is more in man than the mere breath of his body" (p. 477). Wherever he stepped there were mysteries and beauty, and very often both at once.

Fortunately, much grandeur remains. There *are* mysteries. There *is* beauty. Most is hidden in the very large and the very small, but a great deal persists in the folded spaces in between. Whenever I look closely and see something clearly for the first time, then, as with Darwin, *wonder* happens.

In my professional life I study why bacteria have different shapes, which bemuses my friends and astounds everyone else. Who cares what bacteria look like? I respond, "The *bacteria* care," which seems to me reason enough. And *why* they care would seem perfectly reasonable to Darwin. Bacteria, like all creatures, use every available tool to increase their odds of surviving, including adopting specific shapes. Long or very tiny bacteria resist being eaten by protozoa, very long ones are not easily washed from the soil, thin cells accumulate nutrients more readily from watery environments, rod-shaped bacteria move more certainly towards food sources, spherical ones may produce more progeny per gram of resource, spiral ones race more quickly through viscous fluids, flat ones expose more surface area to light, cells just the right size float at just the right depth in lakes and oceans, and triangular cells fit together like so many slices of pie. And yet, all we know or think we know is dwarfed by what we don't. We don't know why some microbes are shaped like six-pointed "ninja" throwing stars, why many look like tiny octopi with multiple arms, why others could pass for microscopic bowling pins, why some coil up like springs or doughnuts, why a few appear as flattened coins or postage stamps, why a number branch in two at their ends, or why one recently discovered bacterium looks like nothing less than a dramatically miniaturized churro (an elongated, deeply fluted Mexican pastry).

What keeps us from being completely bewildered by this confusion?

Charles Darwin.

Thanks to Darwin (and, to be fair, to many of his contemporaries) we have a tool, an outline we can follow to fill in the gaps. We have a frame in which we may place new pieces of information as we work to finish the whole picture, like a jigsaw puzzle with the edge pieces connected. There are two keys: all bacteria adapt themselves to their environment, and all are related.

The first key, adaptation, is commonly referred to as natural selection. All this means is that, as in the Olympics, those bacteria that outcompete the others survive. The playing field and the rules of the game are set by the environment, but within those rules the bacteria are free to adopt any strategy that gives them an advantage. The examples described above hint at reasons why bacteria take on different morphologies. Some shapes are just better suited to getting around in or surviving within a particular environment. One clue that this is so is that each species generally exhibits a single uniform cell shape, suggesting that maintaining a specific morphology gives them an advantage. Another clue is that some bacteria change shapes in response to changes in their environment or during the course of infection. To do so, cells must invest quite a bit of genetic and biochemical effort, and one reason they do is because different shapes are useful in different situations. By learning why bacteria have different shapes, we begin to learn about the physical, chemical, and biological forces that they defy and cope with.

The second key is that all bacteria are related. Just like your nose or earlobe reminds people of your great-aunt or grandfather, the characteristics of bacteria can be mapped onto their family tree. This is a great help in unraveling the history of cell morphology, because at least some of the differences in bacterial shape track along family lines. Scientists used to think, by virtue of just thinking, that the first bacterium must have been spherical, because that's the simplest shape we can think of. And it is, on the face of it. A sphere is just a bag that expands as it grows, like an inflating balloon. But this reasoning leads us astray. By mapping cell shape to the bacterial family tree, it becomes clear that spherical cells (cocci) appear only at the tips of the branches, at the end of family lines but never at the beginning. In fact, spherical cells appear to be dead ends, because once a family line becomes coccoid, no later generation returns to being a rod. So it looks like spherical cells are degenerate forms, not primordial ones. This historical trend tells us that the earliest bacterium was probably a long filament or a rod and gives us a place to start thinking about how bacteria developed. Unfortunately, so far, the more complicated microbial shapes do not follow a clear pattern on the bacterial evolutionary tree, so we don't yet have a good picture of how these other morphologies are related. But, as heirs of Darwin's insights, we are confident that they *are* related and that we can figure out how.

These two Darwinian keys, environmental interaction and family relationships, help explain why bacteria may be one shape rather than another. However, they don't directly address the question of *how* bacteria create a specific shape. Interest in this mechanical question has grown rapidly in the past few years. The combined work of many laboratories shows that all bacteria share a collection of proteins responsible for shaping cells. Some function as internal skeleton-like scaffolds that organize the others in just the right way to build a rigid wall around the cell in just the right shape. My lab has mutated many of these in the common bacterium *Escherichia coli*, which has a boring rod shape (like a small hot dog), and some mutants have "forgotten" how to be a rod. Instead, individual cells are gloriously weird and highly aberrant, so that a single culture represents almost every known bacterial shape. So far the morphologies are random, and we haven't isolated a mutant that grows with just one particular new shape. But the fact that our mutants are exploring the "shape universe" implies that a bacterium can adopt any shape. All that is needed is a mechanism to capture a specific shape and fix it in place. As the community of researchers identifies and investigates these mechanisms, we will, once again, be following in Darwin's footsteps as we complete our understanding of how and why these came to exist.

The young Charles Darwin literally turned over rocks to find and study tiny barnacles. The old Charles Darwin repeated the process in his research on earthworms. Other than just the romantic notion of sailing around the world on the *Beagle* and seeing all things new, I trace much of my Darwin envy to something gradual and subtle, but in its way even more impressive, nestled away in the trajectory of his work. It excites me to see Darwin move, over many years, from gathering a mountain of information about organisms with seemingly nothing in common to the point where *it all begins to make sense*. Every organism is related, in a very particular way, and these relationships can be mapped. "[By considering] descent with modification, all the great facts in Morphology become intelligible" (Darwin, *On the Origin of Species*). This moment of intelligibility is the core of any scientist's life, the reason many of us do science at all. We crave the "high" that comes, all too seldom, when our experiments and thinking fall into place, when everything begins to "make sense." There is simply nothing like it.

Darwin's name is invoked as a great divide between clans of people I like and admire. After 150 years of argument and emotion, we forget that at the beginning of his voyage he was a 22-year-old embarking on a trek around the world, which, though still a feat in our own time, was in Darwin's day majestic. Along the way, he was sick, he was bored, he was enchanted, he was revolted, he was lucky, he was doggèd, he was observant, and he was methodical. The world would change him, and he, the world. In a very small way, that's what I'd like to do.

That you can tell my mother.

FURTHER READING

Amigoni D. 1997. Introduction. *In* T. Griffith (ed.), *Darwin: The Voyage of the* Beagle. Wordsworth Editions Ltd., Cumberland House, Ware, Hertfordshire, United Kingdom.

Griffith T (ed). 1997. *Darwin: The Voyage of the* Beagle. Wordsworth Editions Ltd., Cumberland House, Ware, Hertfordshire, United Kingdom.

Young KD. 2006. The selective value of bacterial shape. *Microbiol Mol Biol Rev* **70:**660–703.

Microbes and Evolution: The World That Darwin Never Saw
Edited by R. Kolter and S. Maloy

doi:10.1128/9781555818470.ch37

37

Postphylogenetics

W. Ford Doolittle

I had the great good fortune to grow up in Urbana, Illinois, a little town with a big university that was the birthplace of our modern understanding of the place of microbes in the evolution of life. As a high school student, I acquired a fascination with microbes from Sol Spiegelman, who gave me a copy of the 1957 edition of Stanier, Doudoroff, and Adelberg's *The Microbial World*. And later in Urbana I contracted an obsession with evolution from Carl Woese, to whom many chapters in this volume could be dedicated.

The Microbial World cautioned me against speculating about the evolutionary relationships of bacteria—not that they didn't have any, but that we would never know them, which jibed well with the fact that no one ever said a word about the subject in any of the classes touching on evolution I had in high school or college. So it seemed to me foolhardy and courageous when Woese undertook, in the late 1960s, to set matters right through an onerous technique called ribosomal oligonucleotide cataloging. What one could do with it, he thought, was collect enough comparative sequence information to quantitate the relatedness between bacterial species and from a table of such pairwise measures construct a universal phylogeny—the tree of life.

Ford Doolittle was educated at Harvard and Stanford and did postdoctoral stints with Sol Spiegelman and Norm Pace. Since 1971 he has been on faculty at Dalhousie University, in Halifax, Nova Scotia, Canada, where he is now Professor Emeritus. He will soon obtain a BFA (Photography) from the Nova Scotia College of Art and Design.

The molecule Carl chose to characterize, 16S ribosomal RNA (rRNA), is ancient, ubiquitous, and essential. It is now the most sequenced of all molecules (or, more precisely, the gene encoding it is the most sequenced of all genes), with 856,341 database entries as I write (May 2009). Nowadays, of course, people can sequence a whole genome (thousand of genes) during lunch, but in the late 1960s it took months to (only very partially) characterize a single rRNA. Also, it took amounts of radioisotopes (^{32}P in particular) that today's students wouldn't even consent to be in the same room with.

My lab in Canada wouldn't have gotten into that tedious but ultimately rewarding business except for the good fortune of being joined by Linda Bonen, who had helped Woese develop the method. With her, we contributed to what I considered cataloging's first major triumph and well worth the effort: proof of the ancient origin from symbiotic cyanobacteria of the organelles which make it possible for algae and plants to fix CO_2 and evolve O_2 (and thus for us animals to live), namely, chloroplasts. Such an origin had been speculated over a century before and promoted in the late 1960s by Lynn Margulis, but rRNA provided the first molecular proof.

My lab then got sidetracked into cyanobacterial molecular physiology, but Woese went on to catalog his way through what seemed at the time a good chunk of the microbial world, and he firmly established the foundations of modern microbial systematics. One of the fruits of this was the discovery of what he then called archaebacteria (now archaea), some of which were already known and acknowledged to be very strange in their biology and ecology. The halobacteria (properly haloarchaea) were a particular Canadian specialty at the time, and so my lab detoured into their physiology and molecular biology, again following Woese's discovery of their evolutionary uniqueness.

It's marvelous how far microbial taxonomy and evolution have come since those early days. Norm Pace's realization that we could amplify (with the polymerase chain reaction [PCR]) rRNA genes directly from environmental DNA samples—and thus address many questions about diversity without the bias and inefficiency inherent in isolation and culture—did for microbial ecology what rRNA cataloging and gene sequencing did for evolution, and brought the two disciplines together as they had never been before. Such "phylotyping" also revealed the vast diversity of the uncultivated microbial majority: most 16S sequences straight from the environment reveal novel and previously uncharacterized "species," and many represent whole phyla of the uncultivated or uncultivatable microbiota. Mitchell Sogin (once Woese's grad student), homing in on a smaller and especially variable region of 16S rRNA and sequencing exhaustively, concludes that microbial communities are vastly richer than generally thought (thousands of taxa meeting operational definitions for species), with much of that diversity rare (a few cells biding their time or just passing through).

As with all good foundational science, the phylogenetic and ecological goals and methods of Woese and Pace have morphed into something more than anyone could have guessed or hoped. And it seems likely that neither the Tree of Life that Woese set out to reconstruct nor the phylotype-based environmental microbiology that Pace founded will survive the first decade of the 21st century with all its paradigms intact.

First, the Tree. In the 1960s, we already knew that genes could be passed between cells of different species of bacteria. Resistance to antibiotics, on the rise in hospitals around the world, was shown by Japanese microbiologists to be due to the transfer between them of small circular DNAs, called plasmids, bearing genes for such resistance. Most of us in the Tree business believed that this phenomenon would be limited in scope and importance outside clinical microbiology: it might even be our fault, a punishment for injudicious use of antibiotics. Certainly that it would be limited was our hope, because if transfer were rampant and affected all kinds of genes (including 16S rRNA genes) in all kinds of bacteria (not just human pathogens), then the Tree of Life would not be recoverable. Different genes would tell different stories, and the Tree would be a thicket of anastomosing branches.

However, if some genes are immune to transfer (as most people still think 16S rRNA genes mostly are), then these might at least record the history of speciation events (which are the branchings in the tree), even though they would be poor predictors of what other genes might share a genome with them. The tree would be festooned with interconnected creepers, but still discernible.

There is continuing debate about the significance of gene exchange, which is called either lateral gene transfer (LGT) or horizontal gene transfer—terms of identical meaning, but like personal computers and Macs, each with its own vigorous defenders. (I prefer LGT and Macs and would be willing to bet there's a correlation.) But nobody denies that most genes have experienced one or more LGT events in their history, and few would claim that even as many as 10% of a contemporary bacterial or archaeal genome's genes have a record of unbroken "vertical descent" (as the lack of LGT is usually termed, though its acronym is not popular). One of many things this means is that even if we could trace the untransferred 10% back to a common ancestral cell that we might call the Last Universal Common Ancestor (LUCA), we would not be able to infer which or how many other genes that entity's genome contained.

Another thing it means is that population genetics, not phylogenetics, may provide the best conceptual foundation for understanding the relationships between microbes. Just as each of us humans is related to all her contemporaries in an uncountable number of different ways, because of the shuffling of genes by recombination—so that there is no single "Tree of Humans"—all bacteria and archaea are interrelated through a complex web of gene exchanges going back four billion years. The genetic processes (and of course the time scale) are different, but it makes no more sense to speak of alphaproteobacteria (say) as a pure lineage than it does to speak of Caucasians as a pure race.

If the Woeseian revolution brought microbiology into scientific "modernity," then the gene exchange paradigm will make it a "postmodern" science, in which fluidity and process are seen as prior to pattern. What this means in less pretentious language can best be illustrated by what metagenomics—into which Pace's phylotyping revolution is now morphing—is doing to our concept of "species." Microbiologists have wrestled with this for decades and are perhaps now about equally divided among those who resolutely believe that species exist, those who think that the idea is nonsense, and those who are hopelessly tired of hearing about it.

As more and more individual strains of recognized "species" have their genomes sequenced (several dozen for *Escherichia coli*, for instance), the more variation in gene content we see: often less than half of a genome comprises genes shared by conspecifics. And the genes that are shared often exhibit high-frequency interstrain homologous recombination and, for that matter, inter-"species" recombination. There are often few barriers to such transfer—there is no reason, in fact, why there should be—and obvious advantages to recruiting potentially useful variants of existing genes from far away in meeting evolutionary challenges.

What all this means, I think, is that we can expect natural environments to contain a "lumpy continuum" of microbial genotypes and phenotypes. The more tightly clustered lumps we might still want to call "species," but there is nothing in the processes that we currently understand that requires (or even necessarily favors) that all cells belong to one of these tighter clusters.

Phylotyping had been until now pursued as if its primary goal were the enumeration and identification of microbial species, as their distributions and abundance are influenced by biotic and abiotic forces. Personally, I think that this whole agenda will have to focus further down (at genes, as they vary in time and space) and further up (at communities, defined as recurring patterns of co-occurring genes)—and forget about species.

This is all good. As Maynard Smith wrote some 30 years ago now:

> The modern synthesis of the 1940s was concerned with eukaryotes.... Its essential achievement was to bring together two previously separate disciplines—the chromosome theory of heredity and the study of natural populations. The same synthesis is now required for the prokaryotes. There is an abundant knowledge of their genetics, but as yet no adequate synthesis of that knowledge with a study of the natural history of bacteria. For example, we have little idea of the significance of conjugation for bacterial populations; it is as if we had no idea of the significance of sexual reproduction for populations of birds and insects. Population thinking has been well developed for fully half a century, but has yet to be adopted by microbiology.

I'd say we are about halfway there.

FURTHER READING

Maynard Smith J (ed). 1982. *Evolution Now.* W. H. Freeman and Co., San Francisco, CA.

Microbes and Evolution: The World That Darwin Never Saw
Edited by R. Kolter and S. Maloy

doi:10.1128/9781555818470.ch38

38

Irreducible Complexity? Not!

David F. Blair and Kelly T. Hughes

Many bacteria have a propeller device called a flagellum on their surface that allows them to swim from one place to another. Bacterial flagella are complicated structures that have been touted by creationists as an example of "irreducible complexity" in an effort to refute evolution by natural selection. The work of many scientists from around the world, including our own labs, has led to an understanding of how this complex structure is assembled from many distinct parts, and how such a complicated structure may have evolved. We will take you on a brief tour through the flagellum assembly manual, and once you've seen the assembly directions, you'll see that instead of refuting Darwin, the flagellum provides convincing support for evolution by natural selection.

Flagella of *Escherichia coli* and *Salmonella* consist of thin helical propellers turned by rotary motors in the cell membrane, thereby allowing the bacteria to move from one place to another in an aqueous environment. This motility provides an enormous survival advantage by allowing bacteria to move toward nutrients or away

David Blair did graduate work at Caltech and postdoctoral training at Harvard before joining the faculty at the University of Utah in 1991. When he's not watching bacteria swim, he enjoys swimming himself, and chasing his daughters down the Alta slopes. *Kelly Hughes* is Professor of Biology at the University of Utah with expertise in microbial genetics. His research interests involve assembly of the bacterial flagellum and coupled gene regulatory mechanisms. He has made fundamental discoveries related to flagellum assembly and general mechanisms of bacterial type III secretion.

from harmful substances. Comparing sequences in bacterial genomes indicates that evolution of flagellum motility was an ancient event. Genes that encode the flagellum occur in about half of the more than 1,000 bacterial genomes that have been sequenced, including those of bacteria that are very distantly related by many other criteria.

The flagellar motor obtains energy from the membrane ion gradient, harnessing the flow of ions from outside to inside the cell to drive rotation of the propeller filament. The flagellar filaments are rigid, helical structures. The flagella turn very rapidly, from several hundred to more than 1,000 revolutions per second. This rapid rotation enables bacteria to swim at speeds of more than 50 μm per second (more than 20 body lengths) in spite of the large viscous drag experienced by objects this small.

The structure of the bacterial flagellum is commonly divided into three parts: (i) a long helical filament, (ii) a hook-shaped structure (called the hook), and (iii) a structure embedded in the membrane that consists of rings mounted on a rod (called the basal body) (Fig. 1B).

The basal body is the motor that turns the flagellum; this motor is surrounded by a set of membrane-embedded protein complexes that form a stator (by analogy with the stator of electrical motors). The rod functions as a driveshaft, transmitting the torque generated near the inner end of the basal body to the hook outside the membrane. The basal body also houses the flagellar export apparatus, which is required for assembly of the structures on the outside of the bacterium. The hook is flexible and thought to function much like a universal joint, allowing torque to be redirected from the motor into the single, coaxially rotating bundle of flagella. The rigid flagellar filament is built from several thousand subunits of the protein flagellin (encoded by the *fliC* gene). The filament, by virtue of its helical shape, converts rotation into linear thrust.

Electron microscopic reconstructions of the basal body show an elaborate structure resembling a chess piece (Fig. 1C). Although these structures look complicated, the MS ring, P ring, and L ring are each composed of single proteins, while the larger C ring is composed of just three different proteins. Further, all of the

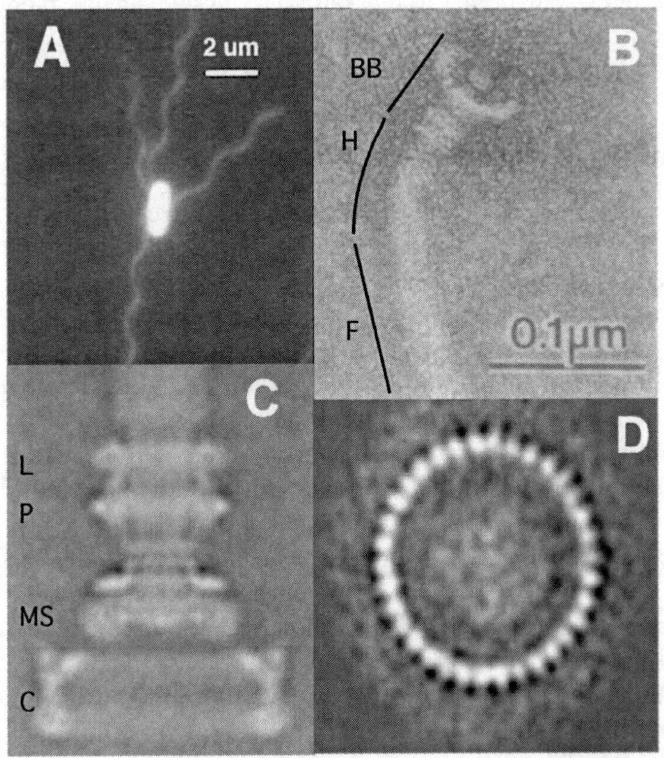

Figure 1 Flagellated bacteria and their ultrastructure. (A) A multiflagellated bacterial cell. (B) Flagellum of *Salmonella enterica* serovar Typhimurium. BB, basal body; H, hook; F, filament. The filament is typically ~10 μm long and only a small part is shown. (C) EM reconstruction of the *Salmonella* basal body at ~22 Å resolution. Basal body rings are named according to their locations relative to cell structures: L, lipopolysaccharide; P, peptidoglycan; MS, cytoplasmic membrane and supramembranous; C, cytoplasmic. (D) Bottom view showing subunit structure in the C ring. doi:10.1128/9781555818470.ch38f1

proteins forming the axial structures (rod, hook, and filament) are homologous, suggesting that they were derived from a common ancestral gene. These protein components self-assemble into their proper positions. Expression of individual proteins or group of proteins can result in the production of the corresponding substructures (MS rings, MS+C rings, or PL rings). A mutant lacking the filament cap protein will secrete flagellin monomers into the extracellular medium. Remarkably, if these flagellin monomers reach a high concentration outside the cell, they will polymerize directly onto the structure in the absence of the filament cap.

Is the Flagellum Irreducibly Complex, or Just Complex?

It is clear that the flagellum is a complex structure and that its assembly and operation depend upon many interdependent components and processes. This complexity has been suggested to pose problems for the theory of evolution; specifically, it has been suggested that the ancestral flagellum could not have provided a significant advantage unless all of the parts were generated simultaneously. Hence, the flagellum has been described as "irreducibly complex," implying that it is impossible or at least very difficult to envision a much simpler, but still useful, ancestral form that would have been the raw material for evolution. This proposition, which is essentially a negative hypothesis, assumes that our present knowledge of the flagellum is complete and that it is impossible to conceive of a credible scenario for evolution of the flagella. In the time since the flagellum was nominated as an example of irreducible complexity, some important evolutionary relationships have come to light which show that the flagellum is not irreducibly complex and which provide the outlines of a credible pathway for flagellar evolution.

The flagellum contains a rotor and a stator; if these are capable of no functions other than flagellar rotation, then both sets of proteins must be present together before any useful function results. However, it has been known for several years that the stator proteins MotA and MotB are homologous to ExbB/ExbD and TolQ/TolR, other membrane-protein complexes that use the proton gradient to energize cellular functions. The Exb proteins function in the active transport of vitamin B_{12} and other essential molecules across the outer membrane, while the Tol proteins function in maintenance of the outer membrane. Studies of their organization indicate that they are very likely organized in essentially the same way as the MotA/MotB proteins, including the presence of an invariant, functionally critical Asp residue that functions in proton translocation. Both the ExB and Tol systems are likely to perform useful functions in the outer membrane by undergoing substantial conformation changes, energized by proton movement through the inner membrane. In this way also, they resemble the flagellar stator, which is believed to be a proton-driven actuator of movement. These connections substantially simplify the problem of flagellar evolution, because they show that an ancestral form of the stator very likely existed and contributed important, proton-energized functions to the cell.

What about the rotor, which is more complex? The flagellum is closely related to the injectisome, which allows pathogens to inject protein cells of an animal host. It is not yet clear whether the flagellum or the injectisome came first or whether both are descended from something else that either has been lost or is present only in niches not yet explored by genome sequencing. In any case, it is clear that export is the essential function carried out by both systems; before the flagellar rotor could have served as a useful organelle for motility, it would have provided an exterior appendage that conferred some useful function, such as (for example) adhesion. (See chapter 2 in this book for another example.)

Such an ancestor need not have been as complex as the present-day flagellar rotor or the injectisome. It would not have required homologs of the three proteins FliH, FliI, and FliJ, which are dispensable for flagellar export. These three proteins could have been added as a later refinement, presumably borrowed from the ATP synthase. Ten other proteins were believed to be essential for flagellar export (FliF, FliG, FliM, FliN, FliO, FliP, FliQ, FliR, FlhA, and FlhB). However, we have recently found that a functional core export apparatus can be formed from just a subset of these (FliP, FliQ, FliR, and FlhA); the others make export more efficient and, again, could have been added to this core in stages. An ancestral rotor would have exported subunits to form some exterior appendage. Because the axial proteins are homologous, and are all capable of self-assembly, we can envision this also occurring in stages, involving duplication and mutation of a single ancestral axial protein.

With an ancestral rotor in place, and ancestors of the stator also present, the stage would be set for evolution of a functional flagellum. In the present-day flagellum, the stator interacts with the rotor protein FliG, which occurs only in flagella and not injectisomes. A key step, therefore, would have been the acquisition of FliG by the rotor. A relevant finding here is that FliG is homologous to the Mg^{2+} transport protein MgtE, which is very widespread in bacteria.

In summary, even given our present, imperfect knowledge of flagellar structure and function, we can outline a credible scenario for the evolution of this organelle from simpler elements. The suggestion that the flagellum is irreducibly complex was essentially a negative proposition; because with our previous, more limited knowledge about flagellar structure and function, scenarios for flagellar evolution were hard to envision, it was proposed that such evolution could not have been possible. We can now say that the proposition was founded on critical gaps in knowledge, particularly a lack of information regarding the evolutionary relationships between flagellar components and components of other systems, and the underlying simplicity in the flagellar export apparatus itself. Thus, rather than debunking Darwin, the bacterial flagellum actually provides further evidence for the evolution of complex structures via natural selection!

FURTHER READING

Berg HC. 2003. The rotary motor of bacterial flagella. *Annu Rev Biochem* **72:**19–54.

Minamino T, Imada K, Namba K. 2008. Mechanisms of type III protein export for bacterial flagellar assembly. *Mol Biosyst* **4:**1105–1115.

Pallen MJ, Gophna U. 2007. Bacterial flagella and type III secretion: case studies in the evolution of complexity. *Genome Dyn* **3:**30–47.

Microbes and Evolution: The World That Darwin Never Saw
Edited by R. Kolter and S. Maloy

doi:10.1128/9781555818470.ch39

39

Many Challenges to Classifying Microbial Species

Stephen Giovannoni

A good microbial ecologist, like my graduate mentor Richard Castenholz, can take you on a walk with a field microscope through Yellowstone National Park or a swamp and point out recognizable populations of microorganisms. Some of these organisms are beautiful, and despite their exotic names—*Sulfurihydrogenibium, Chloroflexus, Oscillatoria,* and *Chromatium*—they fit into the ecology of the landscape in important ways that we know and understand. In fact, microbial populations exist nearly everywhere; and microbiologists with expert knowledge of environments as alien as a deep-sea hydrothermal vent or as familiar as human skin can easily do what my mentor did. By growing colonies of microbial cells on an agar plate, they can name some of the species correctly by eye and get most of them right with a few biochemical tests. There's no denying, then, that there exist things in nature that look and act like microbial species. The familiar way we name and relate to large organisms can work for microorganisms—but only sometimes.

Stephen Giovannoni obtained an M.S degree from Boston University, working with Lynn Margulis, before moving to University of Oregon for a Ph.D. Now a professor at Oregon State University, he enjoys surfing, sailing, and soccer. For decades he has spent part of each summer at the Bermuda Institute of Ocean Sciences, studying the ecology of ocean gyres.

Nature walks aside, studying the natural history of microorganisms is uniquely challenging for a host of reasons, one of which is that there are many different microbial "species"—or at least groupings that look and act like species. I use quotation marks because microbiologists still disagree as to whether microbial species exist, although all agree that there are a lot of different microorganisms on Earth. Of course, there are a lot of beetles too, which prompted a famous quip by J. B. S. Haldane. When asked, "What has the study of biology taught you about the Creator?" Haldane replied, "I'm not sure, but He seems to be inordinately fond of beetles." If Haldane had known what we know today about microbial diversity, he surely would have said "bacteria" instead, or perhaps "viruses." But, diverse as they are, naming beetles as species does not present fundamental problems; it only requires thorough biological procedures and perhaps a particular cleverness with Latin and Greek, the languages used for most scientific names.

So why does the concept of bacterial species evoke such controversy, and why does this controversy increase even as genome sequencing reveals the intricate details of microbial cells? Part of the problem is that no two microbial cells taken from nature are exactly alike. If ever there was a doubt, metagenomics—the process of sequencing bulk DNA from nature—has made it clear that microbial cells are very diverse at a genetic level. I came to grips with this the very first time I saw metagenomic data in the form of genes taken from seawater to identify marine bacteria. We were fascinated to discover that although similar genes clustered to form groups, no two were identical. Apparently, naming species with metagenomic data was not going to be so easy. Of course, this difficulty is not solely the province of microbiologists; if you open the drawer labeled "English Sparrows" in a museum of natural history, you'll see that these birds, after they were introduced to North America, diversified into tens, if not hundreds, of different morphotypes, many of which could easily fool an untrained observer into thinking they were different bird species. What is different about the microbial world?

Other branches of biology have encountered problems similar to those that plague the microbiologist and have dealt with them, but the problem with microbial species isn't merely that there are many of them, or that there is so much variation. What makes microbial species especially difficult to define is the very complicated nature of microbial genetics, which has defied the enduring efforts of humans to explain things with simple models. Wikipedia provides a standard definition that highlights the difficulties faced by microbiologists: a species is "a group of organisms capable of interbreeding and producing fertile offspring, and separated from other such groups with which interbreeding does not (normally) happen." Strictly speaking, this definition doesn't apply to bacteria because they don't have sexual reproduction in the usual sense of the term. Bacterial and archaeal cells don't pair, recombine gametes, and undergo meiosis.

But just as bacteria in nature seem to a trained eye to form recognizable species, they also appear to have something similar to sex when scrutinized from a genetic perspective. In fact, this genetic exchange among bacteria is rampant in nature. Sex is basically recombination generating variation, which is then acted on by selection to produce evolution. In place of sex, bacteria have conjugation, transduction, and transformation. None of these processes is as efficient as sex in eukaryotes, in the sense that real sex is a process that randomly distributes variation within an interbreeding population. However, microorganisms do have a distinct edge in one respect: while eukaryotes stick to closely related partners, bacteria can sample genes from nearly any source, and if the new DNA is valuable to the cell, selection seals the deal. As a result, microbial genomes are a patchwork quilt of DNA from different evolutionary sources.

When microbiologists first became aware of microbial promiscuity, some proposed that the species concept should be retired from microbiology, but there is no sign that such a change will come about. In fact, microbiologists still use the same binomial system to name bacteria that Linneaus devised. The first reason for this is practical: we have to name things in order to organize knowledge. The second reason is more meaningful: there's no denying that in nature, genes are organized into microorganisms to produce patterns that look like species, regardless of whether the process resembles sexual reproduction. Thus, populations of *Thiothrix*, swaying in the sulfidic waters of a Yellowstone hot spring like a lion's sun-bleached mane in the wind, are unmistakable. But the question remains: how do microbial species manage to exist when the familiar mechanisms of sex (meiotic recombination) are not at work to harmonize a pool of genes, and we instead find cells that are varied and cosmopolitan collections of genes; where some genes are in transit, assisted by mechanisms that transport itinerant genes from opportunity to opportunity; some were parked by accident during a passing blight; and some, once useful, are now marginally supported by selection? Microbiologists haven't agreed on an answer to this question, but they are sifting through data to synthesize a new theory of microbial species and diversity that brings Darwin's views of natural selection into alignment with modern views of microbial genetics and global ecology.

So far, we've covered some basics: (i) for those who have never been on a microbial nature walk, there are some amazing, very recognizable microbial species; and (ii) contrasted with the efficiency of meiotic sex, the microbial version of genetics sounds like it was modeled on Grand Central Station. Two additional observations are critical to understanding the arguments about microbial speciation. The first is the importance of the vast size of microbial populations in nature, and the second is the subtler concept of neutral variation and sequence space.

Microbial populations vastly outnumber populations of multicellular organisms, particularly the charismatic fauna and flora we admire. I'm fond of gazing across the sea surface in mid-ocean and contemplating how many of my favorite organism—the bacterium SAR11—are within eyesight. These cells sometimes reach numbers of 1 billion per liter of water and rarely drop below 1/10 of that. The largest microbial populations—including SAR11—may reach 10^{27}, but 10^9 is closer to the limit for successful animals. As R. A. Fisher showed in the 1930s, and M. Kimura in the 1980s, the size of microbial populations can change the game. The bigger the population, the more efficient evolution is.

Sex isn't everything: it may generate variation efficiently, but the other big factor is selection, where bacteria have an edge with their mind-boggling numbers. Such numbers mean that selection can distinguish between very minor differences in populations of microbial cells and increase the frequency of characteristics that it couldn't act on in small populations of multicellular organisms. Michael Lynch pointed out that this argument may explain why self-propagating "selfish DNA" elements, like introns, are so common in multicellular organisms and so rare in large microbial populations.

The next issue to consider is perfection. In the industrial world this is an ideal, and perfect things are exactly alike; but perfect things in the biological world are rarely exactly alike. Every biologist who works with DNA instinctively knows this because of a computer tool called BLAST, which identifies genes by finding a match. The first genome of an archaeon to be sequenced was that of *Methanococcus jannaschii,* which lives in hot springs on the sea floor and can grow on a mixture of hydrogen, CO_2, and salts. Because this organism has DNA, it must also have an enzyme to make the DNA component thymidylate, but searches of the genome using BLAST failed to turn up any genes that matched known thymidylate synthase genes in the *M. jannaschii* genome. Eventually, using some clever tricks, Rajeev Aurora and George Rose found the missing gene, which was only 7% identical to known thymidylate synthase genes.

This scenario raises a conundrum: if microbial evolution is so efficient, how can vitally important genes be so different in the amino acid sequences to which they correspond? The explanation is neutral variation. In the biological world, structures that are nearly perfect in function can look very, very different. These sorts of differences are more apt to accumulate in very large, very old populations of organisms, which describes microbial populations quite accurately.

Microbes are the frontier beneath our feet, yet it has taken a long time for Darwin's ideas to approach their full potential for interpreting microbial diversity. Natural selection is probably very efficient at improving microorganisms to make them nearly ideal "machines" for reproducing in natural environments, but at the same time the age and size of these populations, the constantly changing environments they must adapt to, and their varied repertoires of DNA acquisition mechanisms make them extraordinarily diverse.

Thus, the problem with the microbial species concept lies in knowing where to draw the line. Microbiologists entered the new millennium with wonderful tools to measure microbial diversity in cultures and in nature, and one of the wonders of technology is that it has made these organisms much more visible, revealing previously unimagined details of their form, function, and diversity. These tools are revealing the incredible extent of microbial diversity, but they don't tell us where one species ends or another begins. Thus, interpreting data presents both major challenges and tremendous opportunities in evolutionary biology, and for those involved in the task of defining microbial species today, numerous options and questions remain.

There are aesthetic qualities to science that lie in the power of truly rational explanations and in the beauty of the organisms we study. I began this essay with a reference to my graduate mentor, Richard Castenholz, who brought a sense of natural history to microbiology for his students by instilling observational craft and instinct. My parting thought is an aesthetic experience that imbued the phrase "microbial species" with a different meaning for me. I'm imagining a very cool and clear Oregon spring at Mares Egg Spring, where colonies of the cyanobacterium *Nostoc* once deceived me into asking myself, "How did those beautifully round stones land with such perfection on that bed of fine gray sand?" The answer is that billions of years of evolution brought them there.

Index

A
Actinobacteria, 74, 184, 189
Adaptability
  horizontal gene transfer and, 88
  human analogy, 87–88
  mutations and, 88–91
Adaptation
  of *Bacillus simplex* from "Evolution Canyon," 225–231
  bacterial shape and, 265
  preadaptation, 100, 102
Adaptive immune system, 259–260
Adaptive radiations, 242–245
Addiction modules, 134
Agrobacterium, megaplasmids of, 137
Agrobacterium tumefaciens, 136
Altruism, 78, 80–81, 83
Amanita muscaria, 251–254
Amigoni, David, 263–264
Amino acids
  essential, 194
  produced by *Buchnera*, 194–195
Amplifications, 127, 128
Animal-microbe associations, 173–180
Antibiotic resistance, 49–58
  β-lactamases, 54
  biofilms and, 79
  current state of, 57–58
  drug-focused mechanisms, 54–55
    drug efflux, 55
    drug inactivation, 54
    drug influx, 55
  fitness increase in resistant bacteria, 52
  genetic pathways for evolution of, 51
  horizontal gene transfer and, 51, 52, 54–57
  in methicillin-resistant *Staphylococcus aureus* (MRSA), 44, 49, 54
  natural, 51
  to polymyxin B, 111–112
  rapid evolution of, 45–47
  "Red Queen effect," 45
  selection for, 52–53
  spread by transmissible plasmids, 133
  in *Streptococcus pneumoniae*, 49, 53
  target-focused mechanisms, 53–54
    drastic alteration of target, 53
    mutation of target, 53
    protection of target, 53–54
    replacement of target, 54
Antibiotics
  altruism-inhibiting, 82–83
  annual production of, 46
  broad-spectrum, 40–41
  described, 50

Antibiotics *(Continued)*
ecological role of, 47
history of, 44, 46
mechanism of action, 50
produced by *Pseudonocardia,* 184, 185, 187
rate of discovery of new, 46–47
role as transcription modulators, 41
spectrum of activity, 51
Antivirulence genes, 237, 238
Ants, symbiosis and, 73–74, 182–189
Aphids, 20, 110, 194–196
Aquiflex aeolicus, 19
Archaea, 6, 33, 61, 256, 257, 270
Aristotle, 263
Arsenic, 210–211, 212
Arsenic trisulfide, 210–211
Artificial breeding, within-species variations and, 85–86
The Art of Scientific Investigation (Beveridge), 71
Asexual reproduction, 140, 159–160
Aurora, Rajeev, 285
Avery, Oswald, 234
Avidians, 15
aws locus, 246–248
Azam, Farooq, 65

B

Bacillus anthracis, 82
Bacillus simplex, 227–231
Bacteria; *see also specific species*
ability to spread, 87
generation time, 160
genome plasticity, 103–104
life cycle phases, 115–116
mutualism, 74–76
phage-driven evolution of microbial hosts, 69
phage-encoded toxin genes, 68
robustness of, 99–102
shapes, 264–267
size of, 160
Bacterial adaptation, 99–106
Bacterial populations, heterogeneity in, 102–103
Bacteriophage, 65–69
chimeras, 66
of *Corynebacterium diphtheriae,* 234
cyanophage, 68
global virome, 66–69
infection of bacterial starter cultures in milk fermentation, 148
integration into host chromosome, 148
metagenomics, 67, 68, 69
number per milliliter of seawater, 65
phage-driven evolution of microbial hosts, 69, 148–149
prophage content of pathogens, 148–149
of *Shigella,* 237–238
toxin genes, 67–68, 234
transduction and, 162
Bacterocins, 38–40
Banded iron formations, 221
Barghoorn, Elso, 219
Base substitutions, 127
Bass-Becking, Lourens, 26, 87
Baumann, Paul, 178
β-Lactamases, 54, 82
β-Lactams, 54
Beveridge, W.I.B., 71
Binary fission, 13, 119, 148
Binomial system, 283
Biodiversity
biogeography of microbial life, 25–30
map of, 6, 7
microbial census, 31–36
Biofilms, 79–81
Biogeochemical cycles, 31
Biogeography
of *Amanita muscaria,* 252–253
Darwin and, 26, 241–242, 252, 255

dispersal and, 252–253, 258
of microbial life, 25–30, 257–259
Bioluminescent bacteria, 175–179
Biomarkers, 213–214
Biosignature, 212
Biotechnology, 211
Bistability, 102–103
BLAST, 285
Bonen, Linda, 270
Borges, Jorge Luis, 203
Botstein, David, 66
Breitbart, Mya, 68
Buchnera, 20–23, 110, 194–196, 239

C

cadA gene, 238
Cadaverine, 238
Cairns, John, 105, 125–127, 129–130
Calyptogena okutanii, 20
Cambrian period, 218, 242, 245
"*Candidatus* Korarchaeum cryptofilum," 19–20
"*Candidatus* Pelagibacter ubique," 19
"*Candidatus* Phytoplasma asteris," 19–20
Casas, Veronica, 68
Castenholz, Richard, 281, 286
Cell, minimal, 17, 18
Cell cycle, bacterial, 101–102
Cell division
binary fission, 119
natural selection favoring, 80
Cell morphology, bacterial, 264–267
Chisholm, Sallie, 68
Chlorophyll *b*, 167
Chloroplasts, ancestry of, 167, 192–193, 270
Christian beliefs, 225–226
Christmas Island, 252–254
Ciprofloxacin, 54
Citrate, metabolism by *E. coli*, 15–16
Classification, 281–296
Coevolution
cheaters and, 200
Darwin's tangled bank and, 188
described, 188
Helicobacter pylori and humans, 197–202
pathogenic species, 233–240
COINS (conjugation inhibitors), 137
Collector curves, 34, 35
Colony
bacterial, 62
"morphotypes," 117–118, 243, 244
Communities, competition and cooperation in bacterial, 38–41
Community assembly, spatial theory of, 29
Competition, 243
Competitive-Exclusion Principle, 74
Complexity, reducible, 17–23
Comte, Auguste, 78
Conjugation, 56, 133–138, 160–161, 283
conjugative plasmids, 133–134
efficiency and promiscuity of, 136
inhibitors of, 137
interkingdom, 136
in *Shigella*, 236–237
"shoot and pump" model, 135
steps in, 135
Copy number
of genes, 127, 128
plasmid, 134
Corynebacterium diphtheriae, 234
Coupling protein, 135
CRISPRs, 69
Cross-feeding, 74–75
Crossover, 152
Cyanobacteria, 220, 221
biomarkers of, 213–214
chloroplast origin and, 270
Cystic fibrosis, 90

D

Dairy microbiology, 148
Dallinger, Rev. William, 10
Darwin, Charles
  on ants and aphids, 73
  on barriers between populations, 257
  biogeography and, 26, 241–242, 252, 255
  on descent with modification, 267
  on disuse, 93
  on divergence, 255
  diversity of life and, 241–242
  endemism and, 242
  fossils and, 218
  Galapagos Islands and, 1, 2, 242
  natural selection, 10, 94, 123–125, 242
  natural variation and, 123
  ongoing nature of evolution, 9–10
  On the Origin of Species by Means of Natural Selection, 9, 10, 26, 72, 73, 78, 93, 123–124, 145–147, 181, 218, 233, 255, 263
  spatial scaling of evolutionary diversity, 28
  on species interaction, 182
  stress-induced mutation, 124–125, 126–127, 130
  "tangled bank," 181, 188, 189
  trade-offs, 62–63
  tree of organisms, 146, 152
  The Voyage of the Beagle, 1, 263–264
  on worker sterility, 78
  Yale Center for British Art 2009 exhibit, 72
Darwin Awards, 205
Darwin's Dangerous Idea: Evolution and the Meaning of Life (Dennett), 203
Davies, Julian, 136
Dehalococcoides sp., 19
Delbrück, Max, 11, 12
Dennett, Dan, 203
Descent with modification, 72, 267
Deubiquitinases, 239, 240
De Varigny, Henri, 10
Differential regulation, impact on bacterial speciation, 109–114
Diguanylate cyclase, 246–248
Diphtheria toxin, 234
Dispersal, biogeography and, 252–253, 258
Diversity
  by adaptive radiations, 242–245
  ecological opportunity and, 243
Divinyl chlorophyll, 167
DNA
  as food, 141–143
  methylation, 152–153
  mismatch repair, MutHLS-mediated, 105, 154–155
  selfish, 285
DNA damage
  from bile salts, 101
  SOS regulon and, 100, 101
DNA polymerases
  alternative, 104–105
  error-prone, 101, 104–105
DNA replication
  blockage by fluoroquinolones, 104
  errors during, 119
  flexibility in, 101–102
DNA sequencing, next generation, 35
DNA topoisomerases, 54, 104
DNA transfer, 141, 153; *see also specific mechanisms*
  as bacterial sex, 160–162
  effect on phylogenetic classification attempts, 162–163
DNA uptake, 141–143
  by *Helicobacter pylori*, 199
Dobzhansky, Theodosius, 45, 225, 233

Dogs, within-species variations in, 85
Domains of life, 61, 256
Dominguez Bello, Maria Gloria, 199
Drug efflux, 55
Drug inactivation, 54
Drug influx, 55
Duplications, 127, 128

E
Earth, age of, 218
Ecological opportunity, 243
Efficiency and power, trade-off between, 63
Efflux systems, 55
Ehrlich, Paul, 234
Endemism, 242
Endosymbionts
  Buchnera, 20–23, 110, 194–196, 239
  mitochondria and chloroplasts, 191–193
  reduced genomes of, 18, 20, 193–194, 239
Enteric bacteria, 109–112
Enterobacteriaceae, 109–110
Epigenetic switching, 102
Epigenetic variation, preadaptation of bacterial populations by, 102
Epithelia, colonization by microbes of, 173–174
Escherichia coli
  cell attachment protein, 90
  colony morphotypes, 117–118
  cross-feeding in, 75
  differences from *Salmonella*, 110
  flagella, 275
  genome size, 118
  mobilome of, 133
  in Muta-Flor capsules, 87
  mutations in, 11–16, 152–153
  "pathovars," 236
  polymorphic populations in selection experiments, 74, 75
  recombination, rate of, 153
  recombination with *Salmonella* genome, 155
  reproductive accuracy, 119
  reproductive rate of, 116
  shape mutants, 266
  spontaneous mutation, rate of, 153
  strain differences in genomes, 153
  urinary tract infections, 90
  variability in ability to cause disease, 86–87
Escovopsis, 184, 185, 187–188
Euprymna scolopes, 175–179
"Everything is everywhere" hypothesis, 26, 27
Evolution
  in action, 9–16
  directed, 207
  macroevolution, 86
  microevolution, 85–91, 225–231
  modular, 66
  mosaic, 66
  in nonbiological systems, 150
  phage-driven evolution of microbial hosts, 69, 148–149
  repeatability of, 245–246
  "replaying the tape," 121, 245
  trade-offs, 62–63
Evolutionary diversity, metrics for quantifying, 28–29
"Evolution Canyon," 227–231
Exb proteins, 278
Experimental Evolution (de Varigny), 10
Extinction, 200, 259
Extremophiles, 257

F
Falkow, Stanley, 235
Fatty acid analysis of cell membranes, 229–230
Finches, Galapagos, 2–3, 62–63, 242
Fisher, R.A., 284

Fission; *see* Binary fission
"Fitness" in microbial world, 62, 63
Flagella
  of *Buchnera aphidicola,* 20–23
  complexity of, 275–280
  motility by, 79
  in pathogenic species, 238
Fluoroquinolones, 104
Food, DNA as, 141–143
Formal, Sam, 236
Fossil record, 13, 218
Fossils, 217–222
  age estimation by radioactive decay, 218
  Darwin and, 218
  microscopic, 218–219
  molecular, 213, 214, 219
  morphological, 212
  stromatolites, 220
  trilobites, 218
Fox, G.E., 33
Fuhlrott, Johann, 145, 146
Fungi
  Amanita muscaria, 251–254
  symbiosis with ants, 73–74, 182–189

G
Galapagos Islands, 1, 2, 63, 242
GASP (growth advantage in stationary phase), 115–121
Gastric cancer, 197
Gause, Georgii, 74
Gause's Law, 74
Gedankenexperiment, 245
Gene-inactivation experiments, 18
Gene mixing, 139–1144
Gene networks, 100, 101
Generation time, 160
Gene reduction, in pathogens, 237, 238, 239
Genes
  copy number changes, 127, 128
  essential, 18, 20
  number in mitochondria, 193, 195–196
  pseudogenes, 94–97, 238
  selfish, 66, 67
Genetic architecture, influence on evolution of, 246, 248–249
Genetic barriers, 153–154, 257
Gene transfer; *see* DNA transfer
Genome
  minimal, 17–23
  phage, 66
  plasticity of bacteria, 103–104
  reduction in endosymbionts, 18, 20, 193–194, 239
  size of symbionts, 195–196
Genomic islands, 136–137
Geobiology, 209–214
Goldenfeld, Nigel, 260
Gould, Steven Jay, 121, 245
Grinsted, John, 135
Growth advantage in stationary phase (GASP), 115–121
Guilds, 38, 39, 41
Gut bacteria, mutualism in, 75

H
Hacker, Jörg, 235
Haeckel, Ernst, 3
Haemophilus influenzae, DNA uptake by, 143
Haldane, J.B.S., 282
Hallucinogenic fungus, 252
Haloarchaea, 270
Hardin, Garrett, 80
Hatful, Graham, 66
Hawaiian bobtail squid (*Euprymna scolopes*), 175–179
Helicase II, 154–155
Helicobacter pylori, 197–202
Hendrix, Roger, 66
Henson, Jim, 49–53
Heterogeneity in bacterial populations, 102–103
History of life, 217–223
Horizontal gene transfer, 162, 272; *see also* Conjugation; Transduction; Transformation

antibiotic resistance, 51, 52, 54–57
bacterial adaptability and, 88
gene acquisition by, 95, 110–113
as mechanism for species formation, 136
of plasmids, 136
rate of, 103
regulation of genes acquired by, 112–113
virulence and, 235
Hornets, 77
Host range, effect of pseudogenes on, 95–97
"Host specificity" genes, 95
Hugenholtz, Phil, 69
Humans
coevolution of *Helicobacter pylori* and humans, 197–202
population growth, 80
Huxley, Julian, 11
Hybridization, 179–180; *see also* Symbiosis
Hybrids, interspecies, 154–155
Hypermutable bacterial lineages, 104

I
Ignicoccus hospitalis, 19
Immunity, 234
Immunity protein, 39
Infectious Multiple Drug Resistance (Falkow), 235
Injectisome, 279, 280
Innate immune defense mechanisms, 238
Insects, social, 77–78
Intelligent design, 23
Introns, 285
Iron formations, banded, 221
Irreducible complexity, 23, 275, 278, 280

J
Jacob, François, 240
Japanese pinecone fish, 177
Joyce, Gerald, 65
"Junk" DNA, 113–114

K
Karl, Dave, 178
kcp locus, 237
Kimura, M., 284
Kingdoms, 3, 4, 256
Kirschner, Denise, 198, 200
Klebsiella, 188
Kolter, Roberto, 116
Kropotkin, Prince Piotr Alexeyevich, 78–79

L
lac duplication, 129–130
Lake Louise, 217
Last Universal Common Ancestor (LUCA), 272
Lateral gene transfer, 272; *see also* Horizontal gene transfer
Lawrence, Jeff, 66
Lazar, Sara, 117
Leaf-cutter ants, 183, 185
Lenski, Richard, 155–157
Leprosy, 235–236
Leucine, 196
Lewis, I.M., 11
LexA repressor, 100
Library analogy, 203–207
Life, minimum requirements for, 17–18
Life cycle phases, bacterial, 115–116
Lipids, 213, 214, 219
Listeria monocytogenes, 90
Long-term stationary phase, 115–116
Low-abundance taxa, 34–35
Lowenthal, David, 214
Luria, Salvador, 11, 12
Lynch, Michael, 285

M
MacLeod, Colin, 234
Macroevolution, 86

Macrophages, 149
survival of bacteria within, 90
Malthus, Reverend Thomas, 72
Mann, Nick, 68
Margulis, Lynn, 270
Mar regulon, 101
Marshall, Barry, 197
Maurelli, Tony, 237
Maynard Smith, John, 273
Maynard-Smith library, 203–207
McCarty, Maclyn, 234
Mendel, Gregor, 145, 146
Messenger RNA (mRNA), 4, 60, 100
Metabolic diversity, 212
Metabolism, microbial, 210–214
Metagenomics, 272, 282
Metchnikof, Elie, 234
Methanococcus aeolicus, 19
Methanococcus jannaschii, 285
Methicillin-resistant *Staphylococcus aureus* (MRSA), 44, 49, 54
Methylation of DNA, 152–153
Meyer, Ernst, 153
Microbes
age of origin of, 212–213
numbers of, 31, 32
Microbial census, 31–36
Microbial diversity, 26–29
enormity of, 36, 256
environmental samples, 33–35
geographic barriers and, 257
natural distribution of, 256
rare biosphere, 35, 36
The Microbial World, 269
Microevolution, 85–91, 225–231
adaptive mutations, 88–91
described, 86
horizontal gene transfer and, 88
Mismatch repair, MutHLS-mediated, 105, 154–155
Mitochondria, 191–193
Mobilome, 133, 138
Modular evolution, 66
Molecular phylogenetics, 33
Monera, 3, 6
Monod, Jacques, 91
Morin, James, 176
Morphotypes, 117–118, 243, 244, 282
Mosaic evolution, 66
MotA/MotB proteins, 278
mRNA (messenger RNA), 4, 60, 100
MRSA (methicillin-resistant *Staphylococcus aureus*), 44, 49, 54
Mueller's ratchet, 97
Muta-Flor, 87
Mutation
adaptive, 88–91
amplifications, 127, 128
of antibiotic target, 53
by Avidians, 15
beneficial, 119–120, 124
biodiversity and, 32
deleterious, 119
from DNA replication error, 119
duplications, 127, 128
effect on protein, 206
fluorescent monitoring of, 151–152
GASP (growth advantage in stationary phase), 118–120
hypermutable bacterial lineages, 104
large-effect, 126, 130
in mitochondrial genes, 193
neutral, 94
origins of, 124
pathoadaptive, 89
pseudogenes, 94–97
random, 94, 125, 126
reading frame changes, 127
in rRNA-encoding gene, 60
small-effect, 127–131
spontaneous, 94
stress-induced, 124–125, 126–127, 130–131
transitions, 127
variation in mutation rates, 104–105

Mutation rate, 159
in *Escherichia coli,* 153
selection effect on, 124, 125, 128–129
for small-effect mutations, 127–128
variation in mutation rates, 104–105
mut deficient strains, 156
MutL, MutS, and MutH, 105, 153–154
Mutualism, 71–76, 182–189; *see also* Symbiosis
long-term stability of associations, 189
obligate, 182
mws locus, 246–248
Mycobacterium leprae, 236
Mycobacterium tuberculosis, 53, 90
Myxococcus xanthus, 14

N
Nash, John, 200
Nash equilibrium, 200–201
Naturalist (Wilson), 1–2
Naturalists, 255–261
Natural selection
cheaters, 200
Darwin and, 10, 94, 123–125, 242
library analogy, 204–205
mutations and, 123–125
pathogenic species origin by means of, 233–240
sexual reproduction and, 140
Nealson, Ken, 177
Negative hypothesis, 278
Neutral mutation, 94
Neutral variation, 285
Nevo, Eviatar, 227
Next generation DNA sequencing technology, 35
Niche specificity, 243
Noise, molecular, 102
Nostoc, 286

O
Ochman, Howard, 110
ompT gene, 237
On the Origin of Species by Means of Natural Selection (Darwin), 72–73, 123–124, 145–147, 233, 263
on ant-aphid interaction, 73
on artificial selection, 10
on biogeography, 26, 255
closing passage, 9
"descent with modification" quote, 267
"disuse" quote, 93
ongoing nature of evolution, 9–10
"tangled bank" quote, 181
on worker sterility, 78
Opportunistic pathogens, 73
Origin of transfer *(oriT),* 134, 135
Orpiment, 210–211
Overlap, bacterial robustness and, 101
Oxygen
ancient Earth and, 221–222
from photosynthesis, 211, 213–214

P
Pace, Norman, 6, 33, 256, 270–271, 272
Parasites
of ant-tended fungus gardens, 184–189
obligate, 99
Parasitism, 72–73
Partition systems, 134
The Past Is a Foreign Country (Lowenthal), 214
Pathogenicity islands, 235, 237
Pathogens
adaptive mutations, 88–91
differences between *Salmonella* and *E. coli,* 110
horizontal gene transfer and, 88
host range of, 95–97

Pathogens *(Continued)*
opportunistic, 73
origin of species by means of natural selection, 233–240
phage-encoded toxin genes, 68
prophage content of, 148–149
trait acquisition by conjugation, 138
PCR for DNA amplification from environmental samples, 33, 35
Penicillin binding proteins (PBPs), 53
Penicillins, 54
Peptic ulcer disease, 197
Perfection, 285
Phage; *see* Bacteriophage
Photobacterium leiognathi, 176
Photosynthesis
appreciation of, 165–166
oxygenic, 211, 213–214
Phylogenetics, 269–273
Phylogenetic trees
Darwin's tree of organisms, 146, 152
Tree of Life, 27, 146, 150, 256, 269, 271–272
Phylogeny, described, 33
Phylotypes, 34, 35
Phylotyping, 271, 273
Phytoplankton
cell size, 166
photosynthesis by, 166
triggering blooms of, 171
Picrophilus torridus, 19
Pigeons, 191
Pilus, 161
Plaque, 79
Plasmids, 56, 271
chromosome integrated, 136
conjugative, 133–138
copy number, 134
mobilizable, 134
origin of transfer *(oriT)*, 134
R388, 135–137
role in adaptation, 103
size of, 134
stability determinants, 134
transmissible, 133–134
virulence, 236–237, 239
Plasticity, bacteria genome, 103–104
Polymerase chain reaction (PCR), for DNA amplification from environmental samples, 33, 35
Polymorphic populations in selection experiments, 74, 75
Polymyxin B, 111–112
Polyunsaturated fatty acids, as inhibitors of conjugation, 137
Population genetics, 226, 272
Power and efficiency, trade-off between, 63
Preadaptation, 100, 102
Privileged niches, 199
Prochlorococcus, 165–171
discovery of, 166–167
diversity of strains, 168, 169, 170
"ecotypes," 168
genome sequence, 168–169
number of genes, 169
number of organisms, 166
P. marinus, 19
pan genome, 169–170
psbA genes, 68
Prophage, 148–149
Proteins
directed evolution, 207
Maynard-Smith library analogy, 204–207
psbA genes, 68
Pseudogenes, 94–97, 238
Pseudomonas aeruginosa, 82, 89–90
Pseudomonas fluorescens, 243–248
Pseudonocardia, 184–187, 189

Q

Quinolone antibiotics, 54
Quorum sensing, 81

R
Radioactive decay, fossil age estimation by, 218
Radioisotopes, 270
Random chance, role in evolution, 93–94
"Rare biosphere," 35, 36
Raymond, Percy, 218
Reading frame changes, 127
RecA recombinase, 154–155
Recombination, 141, 272, 273
fluorescent monitoring of, 151–152
inter-"species," 273
interstrain, 273
meiotic, 284
prevention of interspecies, 154
rate in *Escherichia coli*, 153
sex as, 283
"Red Queen effect," 45
Reducible complexity, 17–23
Redundancy, bacterial robustness and, 101
Regulator proteins, 111–113
Regulatory circuits, 100, 111–112
Regulatory RNAs, 114
Reichelt, John, 178
Relative abundance, 34
Relaxase, 135
Reproductive isolation, 153–154
Resistance; *see also* Antibiotic resistance
to bacterocins, 39–40
Restriction endonucleases, strain-specific, 199
Rhizobium, megaplasmids of, 137
Ribosomal oligonucleotide cataloging, 269
Ribosomal RNA (rRNA), 4–5, 60; *see also* rRNA gene sequences
Ribosomes, 4–5, 60, 63
Rifampin, 53
RNA
messenger RNA (mRNA), 4, 60, 100
regulatory RNAs, 114
ribosomal RNA (rRNA), 4–5, 60
RNA polymerase gene, mutation in, 53
Roller derby, 25, 29–30
Rose, George, 285
Roseophage SIO1, 66
Roux, Emile, 234
RpoS-dependent stress response, 104–105
rRNA gene sequences, 270
of environmental samples, 33–35, 270–271
evolutionary comparisons from, 60–61
expense of sequencing, 34
molecular phylogenetics, 33
next generation DNA sequencing technology, 35
sequence alignment, 5
Ruby, Ned, 176

S
Saccharomyces cerevisiae, 155
Saccharomyces paradoxus, 155
Salmonella enterica, 82
differences from *E.coli*, 110
DNA repair in, 101
flagella, 275, 277
Mar regulon, 101
recombination with *Escherichia* genome, 155
serovar Typhi, 94–97
serovar Typhimurium, 94–97, 277
SAR11, 284
Sedimentary rocks, 217, 219–221
Segall, Anca, 66
Selection
effect on mutation rate, 124, 125, 128–129
nonlethal selection regime, 126
SeqA protein, 152
Sequence alignment, of ribosomal RNA (rRNA) genes, 5
Serotypes, evolution of, 238
Sex, bacterial, 133, 139–144, 146–147, 159–163

Sexual reproduction, 139–144, 283
as motor for genetic diversity in a population, 147
Shapes, bacterial, 264–267
Shigella, 87, 236–239
Siderophores, 38
Silencing, 113
Small-effect mutations, 127–131
Smith, John Maynard, 200
Social insects, 77–78
Sociality in microbes, 78–83
Sogin, Mitchell, 271
SOS regulon/response, 100, 101, 104
SoxRS regulon, 101
Spatial distribution of biodiversity, 25–26, 28–29
Spatial theory of community assembly, 29
Speciation, 151–157, 284
of *Bacillus simplex* from "Evolution Canyon," 225–231
horizontal gene transfer and, 136
impact of differential regulation on, 109–114
Species
biological definition, 147, 283
challenges to classifying microbial, 281–286
diversity with bacterial, 147
interspecies reproductive barriers, 153–154
nature of bacterial, 55–56, 57
number of named bacterial, 147
Species-area relationship, 27
Species concept, sexuality linked to, 147
Species richness, 27–28
Spencer, Herbert, 78
Spiegelman, Sol, 269
Squid-*Vibrio* symbiosis, 175–179
Stars, 31
Stewart, William, 44
Streptococcus pneumoniae, 49, 53
Streptococcus pyogenes, 148–149
Streptococcus thermophilus, 148
Stress-induced mutation, 124–125, 126–127, 130–131
Stress response, RpoS-dependent, 104–105
Stromatolites, 220
Sulfolobus, 257–260
Sulfur, 213, 220, 221, 222
Sullivan, Matt, 68
Symbiosis, 173–180
of *Amanita muscaria*, 252
ant-microbe, 182–189
aphid-*Buchnera*, 20, 110, 194–196
evolution and, 191–196
as evolutionary innovation, 189
gene loss and, 193–194
genome size of symbionts, 195–196
mitochondria and chloroplasts, 191–193
squid-*Vibrio*, 175–179
Synechococcus, 166, 167
Systematics, 270

T

Taylor, John, 257
Thiothrix, 283–294
Thymidylate synthase gene, 285
Tol proteins, 278
Toxin, phage-encoded, 68
Toxin-antitoxin systems, 134
Trade-offs, 62–63
"Tragedy of the commons," 80
Transcription factors, 111–113
Transduction, 56, 162, 283
Transformation, 56, 162, 283
Transitions, 127
Translation, 63
Tree of Life, 27, 146, 150, 256, 269, 271–272
Trilobites, 218
Tryptophan, 195, 196

Tuberculosis, 53, 90
Tyler, Stanley, 219
Type III secretion system, 237–240
Type IV secretion system, 135, 137
Typhoid fever, 94, 235

U
Universal tree of life, 5
Ureaplasma urealyticum, 20
UvrD, 154–155

V
vacA gene, 199
Vancomycin, 54
Variation, heritable, 152
Vibrio cholerae, 82
  chromosome 2 of, 137
Vibrio fischeri, 175–179
Virchow, Rudolf, 146
"Virome," 66–69
Virulence genes, 110
Virulence plasmids, 236–237, 239
Viruses; *see also* Phage
  absence from tree of life, 150
  global virome, 66–69
  infecting *Prochlorococcus,* 170
  infection of bacterial starter cultures in milk fermentation, 148
  metagenomics, 67, 68, 69
  number in biosphere, 73
  as obligate parasites, 73
  of *Sulfolobus,* 259–260
  uncultured, 67
von Behring, Emil, 234
The Voyage of the Beagle (Darwin), 1, 263–264

W
Wallace's line, 253
Warren, Robin, 197
Wilson, Edward O., *Naturalist,* 1–2
Woese, Carl, 3–5, 6, 12, 32, 33, 60–61, 165, 256, 260, 269–271
wsp locus, 246–248

Y
Yeast, interspecies hybrids, 155
Yersin, Alexandre, 234
Young, Dick, 176, 178